D0260884

Dog Story
An Anthology

Dog Story
An Anthology

Life and death of our best friends

Anna Pasternak
Jackie Stewart
Petronella Wyatt
Roy Hattersley
Annabel Goldsmith
Philip Treacy
Edward du Cann
Tom Rubython

Press

First published in Great Britain in 2010
by The Myrtle Press

1 3 5 7 9 10 8 6 4 2

A CIP catalogue record for this book is available
from the British Library.

ISBN: 978-0-9565656-1-7

Typeset in Garamond by CBA Harlestone
Reproduction by Fresh Vision, London

Printed and bound in the UK by
CPI Clowes, Ellough, Beccles, Suffolk,
NR34 7TL, United Kingdom

The Myrtle Press
Kemp House
152-160 City Road
London
EC1V 2NX
Tel: 020 7566 1196

IV

for Wilfred
1989 - 2009

for Boss
1990 - 2005

for Mimi
1996 - 2009

for Buster
1995 - 2010

for Copper
1983 - 1998

for Mr Pig
1992 - 2004

for Juliet
1986 - 2004

for Harry
2002 - 2010

'Be comforted, little dog, thou too in the
Resurrection shalt have a tail of gold.'
- Martin Luther

CONTENTS

FOREWORD

ACKNOWLEDGEMENTS

CHAPTER 1.
WILFRED
A love like this
Anna Pasternak
1.

CHAPTER 2.
BOSS
The soul is not easily defined
Jackie Stewart
15.

CHAPTER 3.
MIMI
The promised kiss of springtime
Petronella Wyatt
37.

CHAPTER 4.
BUSTER
Roy Hattersley
53.

'You think dogs will not be in heaven?
I tell you, they will be there before any of us.'
- Robert Louis Stevenson

CONTENTS

CHAPTER 5.

COPPER

The greatest character I ever knew

Annabel Goldsmith

65.

CHAPTER 6.

MR PIG

Not a dog at all

Philip Treacy

77.

CHAPTER 7.

JULIET

Juliet and me

Sir Edward du Cann

99.

CHAPTER 8.

HARRY

Harry's gone to heaven

Tom Rubython

113.

APPENDIX

A History of Dogs

141.

DOG STORY

List of illustrations

SECTION 1
Peter Egan with Fynn, Megan, DJ and Cassie
Wilfred on the Isle of Iona
Wilfred and Daisy Pasternak
Wilfred and Anna Pasternak at Westonbirt
Anna Pasternak and Wilfred on her wedding day
Wilfred and two-month-old Daisy Pasternak
Wilfred and his brother William
Boss and Bugsy with the Stewart family
A portrait of Boss
Sir Jackie and Lady Helen with Boss and Bugsy in Scotland
Boss with Bugsy
Sir Jackie Stewart and Boss on their last day together
Minnie and Petronella Wyatt
Mimi and Verushka Wyatt
Roy Hattersley with Buster and friends in the woods
Buster and Roy Hattersley at the Edinburgh Book Festival

TEXT SECTIONS
CHAPTER 1.
Wilfred begging for it
CHAPTER 2.
Sir Jackie's wee man, Boss
CHAPTER 3.
Petronella Wyatt's Mimi
CHAPTER 4.
Roy Hattersley's Buster

List of illustrations continued

SECTION 2
Mr Pig with Philip Treacy
Copper with Jemima Goldsmith in the swimming pool
Lady Annabel Goldsmith with Copper
Copper in the swimming pool
Mr Pig with Amanda Harlech, Daphne Guinness and
Isabella Blow on a photoshoot
Mr Pig with Philip Treacy and Grace Jones
A portrait of Juliet
Harry, aged 16 weeks, sipping tea
Harry with mother and father, Buttons and Dino
Harry's fifth birthday party
Harry asleep in a filing box
At the seaside with friends
Harry swimming in the engine pond at Castle Ashby
Harry with Daisy and Doris at Christmas
A painting of Juliet commissioned by Sir Edward du Cann

TEXT SECTIONS
CHAPTER 5.
Copper in his mafia outfit
CHAPTER 6.
Philip Treacy's Mr Pig
CHAPTER 7.
Sir Edward du Cann's Juliet
CHAPTER 8.
Harry the Hound in the snow at Harlestone, 2008

Peter Egan and Megan

FOREWORD
Love at first bite

PETER EGAN

I think there is always one dog in your life that is the
gatekeeper; the one who opens your heart to others.
For me, it is DJ, my spollie; our nickname for our
lovely Spaniel/Collie cross. We call him DJ because
he has a white bib and a black coat, and appears
to be wearing a dinner jacket. I found him under a
bucket in the cattery of a rescue centre. He was too
young to be in the main dog kennel and was being
cared for by two lovely female cats, called Franca
and Edna. Seeing his tail sticking out from under
a bucket, I went to investigate. When my face got

too close to his, he attached himself by his teeth to a beard I had grown at the time, and, I suppose you might say, it was love at first bite.

It may have been that he reminded me of a dog I had as a child, or perhaps it was just his wonderful courage and exuberance, but I knew that I had to have this creature. Our home is currently occupied by, in order of rescue, an elderly blond Labrador called Sam, our spollie DJ, a Lurcher Whippet called Fynn, a black Labrador called Cassie and a beautiful Staffordshire called Megan. After Sam, DJ was our second rescue, followed quickly by Fynn, Tucker, Cassie and finally Megan. Megan is two years old and had been used as a breeder. After three litters, she was dumped and ended up on death row in a pound. She was rescued by my charity, All Dogs Matter.

We lost Tucker to epilepsy three months ago. He was the most beautiful creature, a Welsh Border Collie crossbreed, and we had the incredible pleasure of his company for eight years. He was diagnosed with epilepsy when he was three and was on medication for five years. Epilepsy in a dog is the same as it is in humans; he would go for long periods without fitting at all, sometimes for five months. In fact, before he died, we thought he might have grown out of it. Then, over a two-day period, he had a series of cluster fits, went into a coma and we lost him. It was heartbreaking.

Being a Border Collie, he was the glue in our pack. On our walks, he would constantly round up the others and, every ten minutes or so, would report back to me and touch my hand with his soft snout. We miss him desperately. Every day when I walk on the Heath, I see him and I know I must honour him and keep his memory alive, no matter how painful it is. I said to someone recently that if there is a magical eternal garden somewhere, and I do most sincerely hope there is, I know that when I breathe my last it won't be too unpleasant because I know that Tucker will run to me and touch my hand with his snout and we will be off.

'Oh how sentimental', exclaimed the person I was talking to. I was rescued by Oscar Wilde: 'Sentimentality is an unearned emotion', and I added 'but sentiment is pure like compassion.' And that is what dogs do for us: they teach us to stop thinking of ourselves; they take away our self-pretence and make us live in the moment. They are wonderful creatures, all of them.

It always amazes me when I hear there are around seven million dogs in Great Britain. It is life-enhancing to know that there are so many individuals, couples and families blessed with the rare insight into life that comes with owning a dog. A dog depends upon you for everything, and it is for life.

And of course the great thing about all dogs is that they relate to you unconditionally. They don't mind if you look a bit ropey or if you haven't showered. In fact, they quite like it, and they are certainly not 'ageist.' They will kiss any old face, or any young one for that matter, with the same commitment and relish. Though perhaps there might be a preference for very young faces because they generally have the remains of a sticky bun or something that requires a dog's special attention.

I think the life of an actor can in many ways be compared to the life of a dog; even if you are very busy and successful, you spend a great deal of your time following trails that lead nowhere and you are constantly waiting for the bone of a job to be thrown in your direction. There is a constant desire to be loved or appreciated and, indeed, accepted by a large and sometimes judgmental public. In a similar way, a dog offers you the being that it is and wants your kindness and support and, hopefully, your love in return.

I have been very lucky to have had at least seven important dogs in my adult life. The first was a black Labrador called Crackers, the second collapsed in front of my wife Myra on the Chiswick High

FOREWORD

Road, right in front of the Ballet Rambert. And before you could say Ninochka, we had an older blond Labrador called Custard. They both were with us through the late eighties and into the nineties, when we moved from West London to North London opposite West Heath. Perhaps it was the proximity to Hampstead Heath and it's great space for pets and people, or perhaps it was my involvement in a variety of animal rescues that led us to the position where we now house five rescue dogs. Crackers and Custard both lived full lives and loved their adventures on the Heath, and, in the fullness of time, they re-homed themselves in the magical eternal garden. The stories included in *Dog Story*, this marvellous anthology, are rare insights and they are described in the most touching, personal and humorous way. They will make you laugh and they will make you cry, and you will remember them with great heart. Sadly, I was born in the year of the dog, in 1946. I say sadly only because I shall be 64 in September and it seems quite a grand age to be; an age when one should have more than grown up. But I think perhaps one of the greatest reasons I love dogs so much is that they, like me, refuse to act their age and continue their 'puppydom' for their entire lives.

Peter Egan
August 2010

DOG STORY

ACKNOWLEDGMENTS

Thank you

TOM RUBYTHON

If you are a dog owner and a dog lover then there can be no greater pleasure than to edit a book like this. It was simply a delight from start to finish.

The idea came to me two years ago, when I read Sir Jackie Stewart's autobiography *Winning Is Not Enough*. Jackie devoted an entire chapter to his dogs and, particularly, to the life and death of Boss, his Norfolk Terrier. He had a special connection with Boss and the story was heart-warming; it was my favourite chapter in the book. The next time I saw Jackie, I suggested that he write a book about dogs. I'm not sure whether he took the comment seriously, but, when the book didn't appear to be forthcoming, I decided to include his story of Boss in a book that I would publish. Luckily, he thought this was a good idea as well. Jackie and his PA Karen Moss worked hard to find me the photos I needed to accompany the story.

I decided to call the book *Dog Story: an anthology* because that is precisely what it is. I then began the long process of finding another seven similar dog stories written by interesting people to make the book a reality.

Lady Annabel Goldsmith was the first to say yes and to allow her story of Copper to be included. Since then, both Lady Annabel and her PA Judith Naish have been very supportive. Roy Hattersley also agreed immediately to allow his story of Buster to be included, and my special thanks goes to his agent Maggie Pearlstine for making it happen.

Philip Treacy was also very keen to tell the story of his Jack Russell, Mr Pig, whom I soon discovered was a very famous dog in the fashion world, with his own fan club in Japan. Alessandra Greco worked hard with Philip to make a marvellous story of Mr Pig's life.

Coincidentally, Petronella Wyatt told the story of her Papillons Mimi and Minnie in the *Mail on Sunday* just as we were searching for stories, and she had just had some lovely photographs taken by Les Wilson.

Finally, Anna Pasternak agreed to allow the story of Wilfred, her Dachshund, to appear in the book. Anna writes beautifully about her dog, and the interaction of Wilfred with her daughter Daisy makes the story very special. Over lunch with Sir Edward du Cann, discussing another book, I told him about *Dog Story* and he proceeded to tell me about Juliet, who became the subject of another chapter.

ACKNOWLEDGMENTS

Searching for someone appropriate to write the foreword to *Dog Story*, I learned that Peter Egan was involved with a charity called All Dogs Matter and it occurred to me that he would be ideal for the task. It would also afford him the opportunity to give his charity some much-needed exposure. Peter is also donating his fee to All Dogs Matter, which makes a big difference to it.

Lastly, my own eldest dog, Harry the Hound, died during the making of this book and I decided to include her story as well.

A book like this is the work of many people, but special thanks goes to my immediate crew: Carly Butler, my PA; Sophia Doe, our picture editor; Ania Grzesik, who designed it; and Kiran Toor for editing the words and struggling with me to get the right balance of warmth, emotion and the correct English. Finally, my gratitude extends to David Peett and Mary Hynes for selling the book to the 2,000 bookshops across Britain and for organising the publicity; without their work nothing happens.

It is difficult to read these stories without both tears and laughter. And I have had my share of each during these lovely few months as *Dog Story* emerged.

Tom Rubython
11th August 2010

Wilfred begging for it in 1996. He was seven at the time.

CHAPTER 1

WILFRED
A love like this

ANNA PASTERNAK

He came to my wedding and he was there through my divorce. He lay his head on my pregnant stomach and welcomed my newborn daughter home from hospital. He was my rock when my daughter's father left me a single mother when she was two, four years ago.

As she grew up, he was always there watching her, looking out for her. If she went too high on the garden swing, he'd look at me with censure as if to say: 'Aren't you being irresponsible letting her do this?'

I loved him far more than my ex-husband or the father of my child. And they knew it.

He was my best friend and companion. He was the other man in my life – and yet he wasn't. He was my dog – my beloved Dachshund, Wilfred. He died, aged 20, last November 2009.

Because he lived so well and so long, becoming less mobile only in the last month due to a cancerous tumour that we were assured gave him no pain, part of me thought he was immortal.

When he suddenly went downhill after breakfast (which, typically, he ate greedily) on Sunday morning and couldn't move, struggling for breath, we called our vet.

Because Wilfred was legendary in the vet's Henley surgery as the oldest dog on their books, Justin, the vet, promised to come to our house when the end was imminent.

When he arrived, we had Wilfred in his basket, in front of the fire, drifting in and out of consciousness, rasping.

My six-year-old daughter, Daisy, stuck three butterfly stickers on his collar so he would fly free and, along with my mother, we lay by him and thanked him through our avalanche of tears for giving us so much joy, such endless happiness.

Justin explained that the cancer was finally shutting his body down and that it could take six to eight hours for him to die naturally, which would be unpleasant for him. It was far kinder to put him down immediately.

I was holding his paw as the injection was administered and he died as he had lived, beautifully and with dignity.

My mother insisted we celebrate his life, so there was a moment of humour when Daisy rushed to our neighbours to announce: 'The vet has just put

WILFRED

Wilfred to sleep. Please come round to celebrate.'

We toasted him with champagne I was too choked to drink and then Justin took him out to his car.

The way he carefully placed his basket on the back seat, as opposed to putting it in the boot, with such love and respect, broke my heart. It was so unreal that I felt like I was in a film, watching myself.

I started running after the car for one more glance of him as Justin pulled away. I couldn't believe – and still can't – that I will never see him again.

It seems so obvious, but until he died I honestly hadn't fully realised that he was the reason why I have been single for four years. And, actually, have never truly committed to a man in my life.

There wasn't a space in my heart because I was so consumed with loving him. Every night for the past four years, while I put my daughter to bed, he would swagger from his basket by the Aga in the kitchen, to his rosy basket in the sitting room to wait for me and for our evening together to begin.

I would light the fire, then lift him onto the sofa and lie there, watching television or reading, while he snuggled against me. Or if I had a TV supper, he would hover next to me, waiting for me to feed him treats from my plate.

This is the first article I can ever remember writing where he wasn't in his basket in my office.

Tears are streaming now, as they did last night on to my plate as I sat alone and ate a lamb chop realising that there was no need to save him the juiciest bits.

One of the worst moments was the morning after he died, as Daisy and I had a daily ritual when we'd ask each other 'Rosy or Aga?', wondering which basket we'd find him in when we went to let him out.

That bleak Monday morning, we held hands as we entered the kitchen and not seeing him there left us clutching each other, sobbing.

Since becoming a single mother, my friends told me that I had become a virtual hermit, but I didn't believe them. Now, I can see they were right. Why slog into London for a party or bad date when I could snuggle up on the sofa in front of Sky+ with Wilfie and save on a babysitter?

I was gaining a reputation as the local eccentric in Henley, as I would push him around town in Daisy's old pram when it became too far for him to walk.

Once, outside a prep school, a woman peered into the pram and shrieked: 'Oh goodness, I thought I was going to find a child in there, not a dog!'

'Oh, I had one of those,' I deadpanned, 'but she was so boring and such hard work so I gave her up for adoption, got a dog instead and have never looked back.' From the horrified look on her face, for one glorious moment, she believed me.

This isn't my first experience of the savage pain of canine death, yet it feels like my first major loss in life. Four years ago, in December 2005, Wilfred's litter brother William died, aged sixteen. It was a fraught time as my daughter's father had left us two months before and I was struggling to cope. On Boxing Day, I went with my mother and Daisy to the South of France to try and escape the gloom of a broken family at Christmas. On the plane, our main concern was that the arrangements for William and Wilfred would go to plan. A friend of my mother's was coming to stay at her house to look after them. William was ailing with longstanding stomach problems, so we left a list of the plethora of pills and pureed food he had to have on an almost hourly basis.

In France, as the sun shone, our spirits started to inch up a little. We were having lunch on the beach, backed up against the sea wall, enjoying the blush of warmth on our wan faces, when my mobile rang. The minute I heard my mother's friend's voice, I knew. 'I'm so sorry,' she said and my heart fell through the floor. 'William is dead.'

Against the exploding shock, I was frantic that Wilfred would die of a broken heart without his brother. As they had never been apart their whole lives, sleeping curled around each other like croissants, how would he survive? We sat, immobilised on the beach, debating whether we should turn around and go straight back home. In the end, we arranged for Wilfred to be taken to stay with some doggie friends and for William's body to be wrapped up in a blanket and delivered to my garage for us to say 'goodbye' to on our return (subzero temperatures in England meant that his body would not decompose). We stayed in France for four days, mainly because we were too depleted to move.

It's almost comical how we got through the days. Because my daughter got upset if my mother cried in front of her, we'd walk for hours pushing Daisy's pram up and down the *Promenade des Anglais*, trying not to crack every time we saw a miniature Dachshund – and believe me, Nice must be the spoilt Dachshund capital of the world, as there are masses trotting smugly along in all manner of coats and accessories.

As soon as Daisy fell asleep, we'd rush to the nearest bench, sit down, clutch each other and weep. Although the days passed in a haze of unhappiness, having Wilfred to return home to made the future seem bearable. I couldn't wait to hug him. Although he mourned his brother desperately, constantly searching for him in the garden, it was as if he saw that he had new responsibilities. Looking after Daisy and me became his priority and he took on a new lease of life. I adored him for that. Amidst chaos and loss, he continued to be the one constant in our life; our trusted life line. Which is why I had no idea how utterly pulverising the grief of his death was going to be.

I'm a trooper. I've been through horrendously unhappy and challenging times: I've been vilified in the press; stabbed in the back by Hollywood power brokers; had an abortion; and feel a complete failure when it comes to relationships.

But for all of my adult life, I've had Wilfred. He was there to cry into his soft fur and to cheer me. He had a peculiar way of talking; a blend of grunting and purring we called 'grunging' and a robust presence.

So I have never experienced such raw pain or felt so agonisingly, hopelessly alone.

Grief is perverse. When you most want to sleep, shattered by emotion, it wakes you in the early hours; the jagged edges of despair pushing through.

I feel both vulnerable – as if a dandelion spore brushing against me would hurt me I am so exposed – yet invulnerable too, as if nothing can further harm me because I'm already in such torment.

When, last week, a publisher rejected a book I've written, it barely registered.

Uncharacteristically unable to pull myself together, I oscillate between numb shock and howling.

I have cried myself – in shops, the hairdresser, the street, endlessly into my pillow – into a disfigured state.

I can't wear make-up because it will be washed away and because I can't fake bravado anymore. And believe me, aged 42, I'm not young or beautiful, so it's not a look I can pull off.

The only source of solace has been the discovery that I am not alone in being alone because I had a pet and not a man.

When I was weeping outside the school gate, a mother told me that she had been single for six years before she met her husband because she had been so happy with her cat.

A sculptress, they worked and lived together in perfect harmony. Another girlfriend in the States, age 45, who has been single for over a decade emailed me this yesterday: 'I am crying as I type this because I know how hard it was when I lost my dog Geldof after almost 16 years in December.

'I think that dogs give us more than any man ever could because men are incapable of unconditional love, and dogs are the spiritual embodiment of it.

'In fact, I know that there have been times when having a dog and loving it has kept me from giving up on my life.'

Jacqueline Bourbon, a transformational coach who

specialises in grief, says: 'Grief is highly individual and so you can't be prescriptive as to how to deal with it.

'For some people, losing a pet is more significant than losing a friend or family member because of the level of attachment. People say "Get over it, it's only a cat", but you can't because the loss is so great.

'You are the third person who has talked to me recently of their deep bereavement over a pet and two were single women for whom the pet was their companion.

'If you use a pet as a substitute for a relationship when a pet dies, you are in a dilemma. Part of you would like another pet to fill that void while another part, if you are self-aware enough, is saying: "I'm going to be brave and in time find a partner, not a pet."

'There is an exact parallel between people who use pets as relationship substitutes and people who serial-hop in relationships and keep repeating the same pattern because they can't bear to deal with their issues or to be alone.'

I'm aware that it's dysfunctional not to need a relationship with a man because you are besotted by a pet.

In my case, it's probably a reaction to my lack of trust in my choices in men. After I was left a single mother, bruised, I gave up on relationships and felt a form of peace.

But if I know one thing about Wilfred's life, it is that he wanted more than anything else for me to be happy.

And amid the desolation and the horrid stillness in the house which no longer feels like a home without him, I know I owe it to him to open my heart to the possibility, later on, of another love.

Jacqueline Bourbon agrees: 'With grief, the big temptation is to shut down and keep your heart blocked, but that will inhibit any relationship later on as the pain will come back to bite you.

'It's incredibly important to feel the pain and stay open, but not to get stuck in grief and wallow in your loss. You need some sort of closing ritual to honour your pet's life.'

So, Wilfred, here's to you, my dearest friend. Thank you for being there through all the tough times and the joyous ones.

Thank you for loving me and understanding me as you did. I miss you desperately. But thank you also for making space in my heart and life, in time, for a new love.

With you no longer beside me but guiding me, hopefully I'll finally find the courage to truly love again – but this time with a man.

WILFRED

Rainbow Bridge (Anon)

By the edge of a wood, at the foot of a hill
Is a lush, green meadow where time stands still,
Where the friends of a man and women do run,
When their time on earth is over and done.

For here, between this world and the next,
Is the place where each beloved creature finds rest.
On this golden land, they wait and they play,
Till the Rainbow Bridge they cross over one day.

No more do they suffer, in pain or in sadness,
For here they are whole, their lives filled with gladness.
Their limbs are restored, their health renewed,
Their bodies are healed, with strength imbued.
They romp through the grasses, without even a care,
Until one day they start, and sniff in the air.
All ears prick forward, eyes dart front and back,
Then all of a sudden, one breaks from the pack.

For just at that instant, their eyes have met;
Together again, both person and pet.
So they run to each other, these friends from long past,
The time of their parting is over at last.

The sadness they felt while they were apart,

DOG STORY

Has turned into joy once more in each heart.
They embrace with a love that will last forever,
And then, side by side, they cross over together.

Regardless of whom I may love in the meantime, I can't wait to meet my darling Wilfred again one day on Rainbow Bridge because there is a part of my heart that will always be his.

DOG STORY

Boss arrived in Sir Jackie Stewart's life on a beautiful summer's day and left on a beautiful spring day 15 years later. When he died, his owner felt part of his own soul went with him.

CHAPTER 2

BOSS
The soul is not easily defined

SIR JACKIE STEWART

The soul is not easily defined but, to me, it feels like something in my core, a mass which is fused with my background and my values. It has nothing to do with money, material success or any form of achievement; it goes much deeper than that.

It relates to places which have meant a lot to me, like the top of a Scottish glen or on the slopes overlooking Loch Lomond, but it also lies in the countryside; since I was young, I have always felt at peace on a moor or in woods, surrounded by wildlife.

In addition, my soul seems profoundly attached to my family and friends, and, without a doubt, to our dogs. When our Norfolk Terrier, Boss, passed away in 2005, I certainly felt a part of my soul went with him, and his loss created a void and a sense of emptiness that has not properly been filled. It is said that you can replenish the soul by keeping in regular contact with whatever means so much to you, and I certainly do get great pleasure from spending time with people and dogs that are close to me, and from just being in the countryside.

The Queen is a knowledgeable handler of dogs, keeping Spaniels and Labradors as well as the famous Corgis, and her skill and experience is widely respected in the gun dog world. In the late 1970s, Helen and I were delighted when Her Majesty gave us a Labrador puppy, fathered by the famous Sandringham Sydney. The Queen likes to name each of the new puppies and she gave this sweet little black bitch the name 'Snare'. A good many months later, over lunch during a shooting day at Windsor, I found myself sitting on one side of the Queen, with Trevor Banks, a horseman from Yorkshire, on the other. Trevor was a great friend, with a fine sense of humour. 'So how's the dog, Jackie?' Her Majesty inquired.

'Very well, Ma'am,' I replied. 'She's really going well.' 'What did I call the dog, Jackie?'

'Well, Ma'am, you may recall you called it Snare, but I have to admit we have changed the name.'

'Oh?' the Queen said. 'Yes,' I said, stumbling over my words, 'er... the children thought "Snare" sounded a little too much like a trap. She is such a sweet dog they wanted to change her name, Ma'am.'

'What have you called it?' 'Blossom, Ma'am.' 'Is the dog healthy?' 'Yes, Ma'am, but she has a slight speech impediment.' 'What do you mean?' 'Well, Ma'am, it doesn't bark quite like other dogs.'

Her Majesty has a keen sense of humour, and I was starting to get the distinct impression that I was being set up. 'So what does it sound like, Jackie?' 'Well, it sounds a bit like a seal, Ma'am.'

'And what does a barking seal sound like, Jackie?'

Imitating Blossom's peculiar bark was not going to be easy, but I had no option but to give it my best shot, and I made a noise that sounded something like: 'Ahwaaah! Ahwaaah!'

Trevor Banks had not been listening to the conversation, but this unusual noise attracted his attention and, ever eager to engage in any banter that would embarrass me, he leaned across, smiling self-confidently at the Queen, and said: 'Knowing Stewart Ma'am, he'll have got the dog cheap from a poor home.'

The entire lunch party erupted in laughter, causing such a commotion that members of the Royal Protection Group rushed in to see what was happening. When the laughter had subsided, the Queen turned to look at Trevor, then turned back to me and, keeping a completely straight face, said: 'Jackie, does Mr Banks know from whom you got the dog?'

That prompted another explosion of laughter. It was wonderful, and Trevor never lived it down.

Even with her speech impediment, Blossom became a

much-loved member of our family. I have always loved animals. Ever since my first day at school, when I found a stray cat on my way home and was allowed to keep it – we called it Fluffy – there had hardly ever been a time when we haven't had a pet in the house. What joy they have brought. What companions they have been. In fact, they have been more than companions. In a frenetic and demanding life, my dogs have been a sanctuary. Far away from the spotlight, removed from the world where you must be properly dressed and carefully prepared, far from any kind of judgement and expectation, they offer unconditional love – and I seem to mean as much to them as they mean to me. The times in my day when I take the dogs for a walk or just get down on the floor and play with them, giving them kisses and cuddles and tickling their tummies, have always been important to me. In some respects, it's a kind of therapy.

Anybody who has felt the remarkable bond that can develop between a human being and a dog will understand precisely what I am talking about. However, I realise some people may be surprised, and a few might even mock the suggestion that these four-legged friends can be so important. I don't mind. Dogs have played a hugely important part in my life, providing countless hours of joy and pleasure. I have cared about them so much and, in one or two cases, their loss has plunged me into the depths of despair and grief. The old saying runs that 'you can judge a person by the way they treat dogs'. It may seem a bit simplistic, but, in my experience, it's not far from the truth. Show me a person who meets a dog for the first time and can put it instantly at ease, and the chances are that individual will be pretty decent.

I got my first dog when I was nine. Prince was a Springer Spaniel, and he became a loyal friend to me throughout my teens and into early adulthood, right through until the day in 1962 when he was off food and seemed a bit off colour. My memory is absolutely clear – I took him to see the vet in Helensburgh, and, standing in the small room, I thought Prince was being given some medicine to make him better. In fact, without saying anything to me, the callous man in the white coat had given him a lethal injection and put him to sleep. I was shocked. At that time, I was twenty three years of age, about to be married and an established international clay pigeon shooting competitor just starting a motor racing career, but I was deeply affected by the loss of Prince and the fact his life had been ended in such a casual and cold manner. Even now, it upsets me.

Helen has always shared my affection for dogs, and, when we moved to live in Switzerland, we had first a wonderful Boxer and then Pekinese. Next, there was the sweet black Labrador bitch from Sandringham, whom we grew to love as Blossom. She always seemed so at home at Clayton House, in Begnins, and became such a gentle presence in the years when our boys were growing up. Blossom lived until she was fourteen and a half, until she developed a tumour. Helen and I took her to an animal hospital in Berne, but we soon found ourselves facing that terrible day, which every dog owner will recognise, when a decision has to be made. We stayed with her right until the end, holding and comforting her while the vet ended her life in the most peaceful way.

By this time, Paul and Mark were studying in America and I felt a couple of Labradors would suit us well, but Helen said she would prefer a lapdog. I wasn't keen on the idea, but she looked through some books and set her heart on a Norfolk Terrier.

'I don't know why you want a wee dog,' I said. 'You can't give them a good clap on the flank like you can with a big dog, and they're always yapping.'

'Well,' she said, 'I suppose I would just like something for me that I could pick up and cuddle.'

The compromise solution was to get a Labrador and a Norfolk Terrier, and to hope they would get along.

I contacted Bill Meldrum at Sandringham and he said that he would look into it. He made some inquiries and came back to let me know he felt sure that one could be made available to us. 'Bugsy' duly arrived as a fully-fledged two-year-old gun dog.

In search of a Norfolk Terrier, I contact John Stubbs, the head gamekeeper at Windsor, and a man with a great knowledge of who was who in the dog world.

'You're in luck,' John said. 'There's a breeder in Rugby who has some puppies.' Since we were living in Switzerland at the time, John suggested he would choose one for us, take it home and get it house-trained. He said we would then be able to pick it up from his house when we were next in the UK.

'That sounds perfect,' I replied.

It was a beautiful summer's day when we eventually drove to Windsor, and I arrived at the house to find the door ajar. I rang the bell and, knowing John and his wife Pat well enough, made my way into the house. In that instant, a small dog appeared, running along a corridor directly towards me, in full flight, barking his head off. This was 'Boss'; from the moment we met, he was a tremendous wee dog, always rushing around, constantly on the move, full of personality.

We made arrangements to take the two new additions to our family back to Switzerland and, realising they would have to be put in a box in the hold if they travelled on a commercial flight, we decided to take them in the Jet Stream 31 that I owned at the time. In years to come, when 'passports for pets' started, we flew in private aircraft whenever we took the dogs.

'The dogs always fly privately,' I used to joke, 'but Helen sometimes flies commercial.'

Bugsy and Boss settled quickly at Clayton House and they had the run of the place once we had installed an elaborate fence around the entire property, even reaching several feet underground because Norfolk Terriers are renowned diggers. Almost immediately, and quite remarkably, it became clear that Helen's dog had become Jackie's new best friend. Boss and I seemed to be glued together. Where I went, he went. Where I slept, he slept. Helen started to say we even began to look the same and behave in the same way. He had such character. At times, he would sit, look up at us and almost seem to talk. Moving his head and opening his mouth, he would make sounds with an intonation that resembled human speech. He wasn't barking and he wasn't asking for food or to go outside, but he was certainly communicating. Maybe he had picked up our mannerisms from watching us. In any event, he would just sit there, chatting away to his heart's content.

In 1996, when it became clear that I needed to be in England to create and develop Stewart Grand Prix, our new Formula One team, Helen and I reached the decision that it was necessary to leave Switzerland for a while to carry out this development. It was a terrible wrench for us both because we had been so happy there, and it was a decision made even more difficult by the knowledge that UK law meant that, upon arrival at Heathrow airport, both Bugsy and Boss would have to spend six months in quarantine.

I was desperately worried how they would cope with living in a confined space with a concrete floor, surrounded on each side and above by metal fencing. At least they were allowed to stay in the same kennel, so they had each other for company. Helen and I visited them twice a week every week – very little took preference in my diary – and, day by day, they and we endured the trial.

It has always seemed to me that, as humans, we have only a basic understanding of a dog's inner senses, and this suspicion was confirmed during our visits to the quarantine kennel. We would be sitting there, playing with Bugsy and Boss, when all of sudden the Rhodesian Ridgeback in the next kennel would start jumping up and down, barking and obviously getting very excited.

'What's all that about?' I would wonder.

Sure enough, about twenty minutes later, the owner arrived for a visit. It was uncanny, but it happened over and over again during the six months. I would happily declare under oath that somehow these dogs were able to know their owner was on his way twenty minutes before his arrival, and that period of time represents a fair distance in a car. Quite amazing. I have no idea how they did it, but they categorically did. Human beings may be quite clever, but we certainly don't understand everything.

The vet who used to do his rounds at the quarantine centre was Martin Watson, a Scot, whose surgeries were not far away in Windsor, Egham and Ascot. He is a complete professional with a wonderful manner, and he would become an important person in our lives.

A long six months eventually passed, and it was pouring with rain on the much-anticipated day when we were able to collect Bugsy and Boss from the kennel and bring them to their new home in Sunningdale, where we were living at that time. We walked the boys to the kennel's car park on the lead, with both dogs stopping at every tree along the grass path. They were so excited to be out and about.

'I'm going to take them for a walk,' I told Helen

when all four of us were in the car.

'In this weather?' she said.

They had suffered long enough and, before we had even gone to the apartment, we drove straight to the golf course. We were not going to be put off by a bit of rain, not even by this torrential downpour. Sunningdale is one of the few major golf courses in England where dogs are permitted to run free, and, as we reached the side of the eleventh fairway, I opened the rear door of the car and let them off the lead. What joy. The dogs burst out of the blocks like 100-metre runners and were just so delighted to be free of confinement. Sprinting in figures of eight, Boss chased Bugsy and Bugsy chased Boss. Helen and I watched in fits of laughter and joy. Four drenched souls eventually returned to our apartment. It was still raining hard, but it had been a memorable moment of supreme happiness.

In the weeks that followed, I watched for signs that the period in quarantine had affected them in any way – and I realised Boss had stopped 'talking'. In fact, he never 'talked' again. However, that aside, they both seemed to have taken everything in their stride and they settled happily in England.

We moved to our current home, the new Clayton House in Buckinghamshire, early in 1999 and again took the precaution of fencing a large area where the dogs would be safe to roam. Boss and Bugsy seemed perfectly happy, and Helen and I treated them and loved them as our children. Day after day, they were a constant source of comfort and happiness for us both.

Bugsy had come to us fully trained by Bill Meldrum, a Field Dog Trials world champion in his own right. Even though I thought my Lab might have forgotten his gun dog skills after six years living in Switzerland, I decided to take him out shooting. It wasn't a complete success. He had no problem with sitting patiently at my side, but, when I sent him out to retrieve a pheasant at the end of a drive, when a Labrador would normally be expected to sprint, Bugsy strutted out with considerable style, found the bird and even stopped for a pee on the way back – a performance of which Bill would not have been enormously proud.

He had become accustomed to his home comforts. Like so many Labradors, he had a mild and gentle nature and, more than anything, a soft mouth. He was wonderfully patient and, when Paul and Mark brought their families over for lunch, our grand-children would climb all over him, pulling and tugging, and he would just lie there and take it; almost seeming to enjoy it.

One day in 2004, after feeding them, I took the

dogs out for a walk, and, as he often did, Bugsy went for a dip in the pond on our way back to the house. Emerging from the water, he suddenly seemed to stagger and fell over on his side. There was blood coming out of his mouth and I thought something had bitten him. He was clearly not in a good way, and I genuinely thought I was going to lose him there and then. I didn't know what to do. I felt lost. I had no phone with me to call for help and I wouldn't leave him. He needed comforting so I lay with him and cuddled close. I thought he would be cold but he wasn't. Boss was looking at us anxiously and I started thinking I would have to carry Bugsy back. However, after fifteen minutes, gaining some strength from somewhere, he managed to stand up. Looking a tad unsteady, he walked slowly back to the house, gradually gaining confidence. I wondered if he had had a stroke, but that would not account for the bleeding from his mouth.

Having dried him off, I called Martin Watson and explained what had happened.

'You had better bring him down here,' he said.

So Bugsy was taken straight down to Ascot, and, when Martin called me, he said it was not good news.

'He's got a tumour in his mouth,' he said.

'So what now?'

'Jackie, we could try and remove it, but, to be

honest, I think the operation would be too cruel for a fourteen-year-old dog.'

'So we have to make a decision?' I asked.

'Yes, I'm afraid so.'

I cancelled whatever I had planned for the rest of the day, and Helen and I drove to the surgery, taking Boss with us. We arrived and realised there was really no option. Once Boss had been taken to Martin's office, Helen and I knelt down beside Bugsy on the floor of this small room and stroked our old Labrador friend as Martin gave him his last injection. He just lay there and died as Helen and I wept.

'Maybe you should bring Boss back,' Martin said when it was all over. 'I think it would be for the best.'

So Boss came back in and, very oddly, he walked right around the body of his departed companion almost without looking at him, staying close to the wall. I picked him up and put him down again, and he did precisely the same thing all over again. He seemed to be distancing himself from the death. Dogs won't immediately mourn a loss, but they will do so four or five days later when they start missing their companions.

Bugsy was cremated and his ashes were buried beneath a white headstone in the garden at Clayton House.

Unsurprisingly, Boss seemed to cling to us even

more in the weeks that followed and he travelled almost everywhere with Helen and me. Into the spring of 2005, this wee man, who had always been so energetic, clearly began to slow down and he reached a stage where he seldom ran and only walked when necessary. I took him to see Martin, and a series of test indicated there was some problem with a valve in his heart. Within a week, he had become very poorly and Martin told me there was nothing more he could do. I said I certainly didn't want Boss to suffer, and Martin kindly offered to come up to Clayton House.

That last morning, Helen and I took Boss for one final, gentle walk. I remember the bluebell forest was looking wonderful and we took some photographs together. It's almost impossible for anybody who has not experienced such moments to appreciate the depths of emotion, but, for what was the fourth time in my life, there was no alternative but to do what was best for Boss, however hard it was for us. Sitting at my desk in my study at home, I held wee Boss in my arms as Martin gave him a sedative. Boss drifted off to sleep and I walked through to the TV room, where we had so often sat together. Around five minutes passed, then Martin gave him the final injection and, lying in my arms, with his head against my chest, Boss peacefully passed away.

It's difficult to explain but, as I have said at the start of this chapter, in that precise moment, I felt as if part of my soul went with him. Martin took him away to be cremated and we later buried his ashes beneath a black headstone, beside Bugsy.

I was profoundly and deeply affected by losing Boss. There were other things happening which made it a difficult period in my life, and my frustrations with the situation at Jaguar Racing and the struggle to save the British Grand Prix at Silverstone did not help, but I simply didn't feel right for six months or more. Helen, Paul and Mark were starting to be seriously worried. 'Come on,' I would urge myself, 'get your act together.' I was meant to be a three-time Formula One world champion, a racing driver who had supposedly shown great courage and mental strength in being able to carry on competing when so many of his friends were being killed, but it was starting to look as if I had been deflated and defeated by the death of a small dog.

Maybe it was just a natural symptom of age: it had occurred to me that people do become more emotional and more easily moved to tears as they get older. Maybe it was just a question of time: when I was a racing driver, my life was so full and consuming that there had almost been no time to stop and mourn. That sounds harsh, but that was the reality –

I was always rushing to the next thing. Whatever the reason, I found it exceptionally hard to recover. The grandchildren would come to visit and they would take flowers to put on the graves of Bugsy and Boss, and I would have to make an excuse not to go with them because I knew that, in all probability, I would simply burst into tears.

I regularly telephoned Bill Meldrum during those difficult days. He understood exactly how I felt and, indeed, had lived through the same experience many times before. One day, he asked me if he could send me a poem. It duly arrived in the post, and it was so touching and, sadly, so real, reflecting the emotions every loving dog owner must face at some stage – not an easy read. With Bill's permission, and with thanks to the author whose name I have unfortunately been unable to find, I include the poem in full here. It is called 'A Dog's Prayer':

If it should be that I grow weak
And pain should keep me from my sleep
Then you must do what must be done
For this last battle cannot be won

You will be sad, I understand
Don't let your grief then stay your hand
For this day more than all the rest
Your love for me must stand the test

We've had so many happy years
What is to come can hold no fears
You'd not want me to suffer so
The time has come, please let me go

Take me where my need they'll tend
And please stay with me till the end
Hold me firm and speak to me
Until my eyes no longer see

I know in time that you will see
The kindness that you did for me
Although my tail its last has waved
From pain and suffering, I've been saved
Please do not grieve, it must be you
Who has this painful thing to do.
We've been so close, we two, these years
Don't let your heart hold back its tears

Boss could have written exactly these words for me.

Time moved on and Helen felt getting a new dog would help, but I wasn't keen because I felt nothing could ever replace what we had lost and that trying would be pointless and unfair.

'Come on,' she said. 'We have to move on.' Her

persistence paid off just over a year later when I agreed to visit the breeder of Norfolk Terriers who had bred Boss. There was apparently a new litter, so we drove up to Rugby to have a look. We arrived to find four little balls of fluff huddled together in a cardboard box, all boys. Our 'look' quickly became two hours of amusement at how these little fellows reacted to us being there. Helen and I had such fun that we decided to buy not one puppy but two because they seemed to get along so well.

'I don't think that's wise,' said the breeder. 'If you take two boys, the likelihood is that as they grow up they will fight a lot. I have a bitch who's going to have a litter soon. Why don't you come back in a couple of months for the second puppy?'

The discussion continued and, in the end, the breeder succumbed. Helen and I made it clear we would like to have both the wee boys together. A month later, when they were ten weeks old, we brought them home and decided to call the larger dog of the pair Whisky and the smaller one Pimms.

They have both become a central part of our lives and we both miss them terribly when we are away. Watching them play together is so enjoyable. They truly love each other and, when they run, it's almost as if they are tied together, side by side; and even now, as he gets older, Pimms still jumps around like a bunny rabbit when he wants Whisky to play with him. Pimms is the one who seems full of life and always looking for trouble. He runs everywhere, with no time to walk. Whisky is more docile, never wanting to stray far from me when I'm home. Both get jealous if one or the other gets a little bit more attention and they'll start pushing in and jumping up to get ahead. They are also great lovers of watching television, sometimes getting quite aggressive when other animals appear and disappear from view, excitedly rushing round behind the television to find out where they have gone. It's a great cabaret.

Norfolk Terriers are renowned for their characters, even among Terriers, and our boys are no exception. They are so affectionate and upbeat. It's as if they wake up every morning and suddenly it's a new day; they are so alert and full of excitement and expectation, wondering what adventures lie in store for them. The world would certainly be a happier place if more humans were to awake with such joy.

DOG STORY

Mimi died showing a splendour of the soul; uncomplaining, resigned, loving. 'When my father died I stood over his coffin dry-eyed. But now I cried increasingly at my loss.'

CHAPTER 3

MIMI
The promised kiss of springtime

PETRONELLA WYATT

Early in 2010, to use that Wodehousian expression, I found myself on the horns of a dilemma. After paying my monthly living expenses, I was amazed to find I had £600 spare to spend on myself.

I yearned for a lambskin Chanel flap bag, though these days £600 doesn't give you much bag for your buck. But it would, I discovered while typing the amount into the internet, buy me a dog.

Ye gods! For most of my life I have been a fur- wearing urban fashionista, with a deep-rooted suspicion of quadrupeds – not to mention shmaltzy books and films, such as *Marley & Me*, extolling the very special relationship between man and dog.

I have, moreover, no truck with animal-rights activists, and feel about vegetarians the way Goering felt about culture (I reach for my Browning).

Yet, a few days later, like one woken from a hypnotist's spell, I found to my astonishment that I was standing in the middle of a Hampshire farmyard handing over £600 in return for a five-month-old Papillon puppy called Minnie.

Why? What in heaven possessed me? So far from being a 'dog person', I have often wanted to throttle the blighters. My first experience of a dog was salutary. Indeed, a psychiatrist might argue that it would account for where I have not got to today.

When I was a child, my father owned a King Charles Spaniel called Minette. Minette was an unpleasant creature with a face like James Cagney. Her two attributes were greed and sloth. Napoleon used to say an army marched on its stomach. Minette trundled on hers, whining for more food and aiming her eyes at me like a pair of gun barrels.

The denouement came when I was 11. A chocolate birthday cake was ordered from Harrods. I remember it vividly, with a sort of wild and bitter regret. It was large enough to serve 14 and had ornate gold and silver piping spelling out my name.

The cake was collected a few hours before tea and placed on the lower shelf of the larder. When it was time to light the candles, my mother found the larder bare. Minette was lying prone on the floor, with vestiges of cake beside her. M'lud, she had eaten it.

I cried like Lord Lundy, but my misery was assuaged by the fact that Minette died soon after. I was not always a nice little girl. Her insides, which could no longer bear her gluttony, rose up and protested. The vet said she had literally exploded.

As I prided myself on being firm but unfair, I extended my hatred of Minette to all dogs. When friends presented their mutts, I recoiled like the prudish wife of a Babylonian monarch who enters the room to find an orgy in full swing.

By my twenties, I had mellowed slightly, chiefly because people had stopped introducing me to their dogs. Then, when I was 26, I went to stay with a girl-friend called Romilly McAlpine, who lived in Venice.

Romilly, who was also an urban fashionista, had, to my surprise, acquired a dog called Klemmie. Yet, with huge feathery ears shaped like butterfly wings, silky white fur and an arched tail, she looked less like a dog than some mythological creature.

Her movements were as fastidious as if she were wearing six-inch heels. She seemed to live on air and water.

This amazing dog was a Papillon, a breed seldom found in England. Papillons have been popular with the French, however; hence their name.

Madame de Pompadour had six, and Marie Antoinette four. Legend has it that she hid them under her skirt to give her courage on the way to her execution.

Klemmie certainly had that Gallic élan. I wanted her with me all the time, including in Harry's Bar on the Grand Canal, where the owner kept a special water bowl for her visits.

Romilly insisted that a Papillon was a perfect dog for me: 'low maintenance' – oh, that magic phrase; easy to house-train; intelligent; and clean. She gave me the name of one of the few breeders in England.

A week later I visited the breeder, or attempted to. Her house had wall-to-wall dogs and she was standing on a ladder looking tearful. Out in the garden were more dogs, in a gargantuan heap. I knelt down to see if I could distinguish one from the other.

Suddenly, a puppy extricated herself and leapt into my lap. She was white with sable markings, possessed of huge black eyes, and those ears. It was as if God had kissed her on the cheek and there she was.

She was two months old and had no name. I took her back on the cold London train. As she shivered, I decided to call her Mimi, after the seamstress heroine of *La Bohème,* whose tiny hand was always frozen.

But what began as a fairytale soon turned sour. Back at my small flat in North London, Mimi peed continuously; it didn't seem possible that such a small dog had so much water in her.

She refused to eat. I rang the breeder who told me Mimi didn't like dog food. What did she like, then? Lobster? For an hour I cooked her a risotto with tiny pieces of salmon, feeling increasingly resentful as I ate a sandwich.

I had been told to make her sleep in a basket in my kitchen. Within five minutes there was a scratching at my bedroom door, followed by whimpers which continued for an hour.

Able to bear it no longer, I let her in. Mimi jumped on my bed. I put her on the floor. And so on it went until Mimi won the skirmish. She used my flat as her personal minefield.

When I was at work, I dreaded what I would find when I got home. I was sleepless and panicky. After four days, I decided to take her back to the breeder. I felt a twinge of guilt, but I had failed to form any sort of attachment to this weapon of mass destruction.

That evening, I took her to my father's house nearby. In one corner of the garden was a ledge with a 15-foot drop to some old concrete paving. Irresponsibly, I let Mimi off her lead.

Suddenly, she was no longer there. I knew where she was. She was lying on the concrete, dead. I was a dog murderess.

Time moved with feet of lead until I could no longer put off looking down. Then I heard a happy bark. It must be her ghost, I thought, delighted to be rid of such a horrible owner.

It wasn't. Mimi was pattering around at the bottom as if falling 15 feet was mere dog's play.

With the use of a ladder I managed to bring her

into the house. I introduced her to my parents, who cooed over her embarrassingly. At this point, Mimi employed a masterly piece of strategy.

She rose on her hind legs, licked my father's hand and looked at him through her thick lashes. 'That dog's not going anywhere,' my father cried, almost sobbing. 'She going to live in this house with us.' He looked at my Hungarian mother. 'Oh my God,' she said.

I visited Mimi often, and this 12-inch-high dog began to grow on me. I watched her transformation from a nervous puppy to a young adult with poise and discretion. That inner grace I had seen was now evident in the sweetest of dispositions.

Soon, I was cancelling dates so I could spend the evening with Mimi, whose intelligence was extraordinary. She not only understood requests made in English but also in Hungarian. Her sensitivity was such that when I was upset she tried to divert me.

When I cried, she licked away my tears. I like to think she made me a better, less selfish person. Being close to a dog is civilising. Raised voices caused Mimi to leave the room, so I kept my temper in check. I also learned, rather late in life, a certain responsibility.

When my parents were on holiday, I moved into their house to look after Mimi. I began to under-

Above: Peter Egan with (left to right) Fynn, Megan, DJ and Cassie, who live together in Hampstead, London, with a blond Labrador called Sam (not pictured).

Above: Wilfred enjoying his holiday on the Scottish island of Iona, 1997.

Above: Wilfred and Daisy, aged 5, both sadly aware it was nearing the end of their time together, September 2009. Wilfred was twenty years old when he died, two months later.

Right: Soulmates: Wilfred and I, happy after walking at Westonbirt in Gloucestershire, 1996.

Left: Me on my wedding day with my true love Wilfred, 11th December 1999. The photograph was taken at Fawley Court, Henley on Thames.

Below: Wilfred, the adoring canine 'nana', checking his two-month-old charge, my daughter Daisy, who was born on 18th October 2003. On the right is an ailing William.

Left: Wilfred allowing his brother William the rare position of top dog, 1997.

Below: From the moment they met in Windsor, Boss was a 'tremendous wee dog', always rushing around, constantly on the move and full of personality.

Above: The Stewart family: Paul, Sir Jackie, Lady Helen and Mark with Boss and Bugsy.

Below: Sir Jackie and Lady Helen Stewart with Boss and Bugsy in Scotland filming for a television programme made by his son Mark, called *The Flying Scot*.

Above: Boss with Bugsy, the Labrador who came from HM Queen Elizabeth II's estate at Sandringham.

Left: Sir Jackie holds Boss, 14 and a half years old, on his last day of life in the woods behind his house in Spring 2005. The bluebells were in full bloom as Boss' owner explored the depths of his emotions in a way that he never had before.

Above: Petronella Wyatt with her new dog Minnie. Minnie has taught her that all dogs are unique. She is even naughtier than Mimi and more excitable.

Above: Mimi, Petronella Wyatt's first Papillon, is pictured with her mother Verushka. Mimi gave Petronella a certain sense of responsibility: 'When I cried, she licked away my tears and I like to think that she made me a better, less selfish person.'

Above: Roy Hattersley on a walk with some friends and their Staffordshire Bull Terriers in the woods. Buster enjoyed his walks right up to the end of his life.

Above: Buster and Roy Hattersley together at the Edinburgh International Book Festival, where he became a favourite with literary old ladies who would bring him treats. Buster was born in February 1995 and lived a happy life for 15 years.

stand why dog owners are so passionate about their pets; though Mimi, who had the carriage of a queen, seemed too dignified for that latter appellation.

She was no snob or sycophant. My father was friendly with Margaret Thatcher, who, with Denis, would dine at my parents' house. Father would send 'the ladies' out of the room so the men could discuss 'serious matters'. Lady Thatcher, naturally, was regarded as an honorary man. Denis went with the ladies.

Outraged one evening, I refused to leave. In a gesture of feminist solidarity, Mimi planted herself next to Lady Thatcher. A panzer division could not have removed her from that room. Lady Thatcher was already making Delphic pronouncements.

While I was too nervous to contradict the goddess, Mimi had no such inhibitions. She cocked her ears and listened intently. When Lady Thatcher attacked selfishness and envy, Mimi woofed approvingly.

But when I was reprimanded for remaining unmarried, Mimi expressed her strong disapproval and peed on Lady Thatcher's shoes. The oracle had been defiled! I awaited a steely response.

Instead, our greatest peacetime Prime Minister quipped: 'The Iron Lady is going to turn rusty.'

Everyone was disarmed by Mimi. When I took her for walks, people stopped me in the street. Initially,

I harboured the forlorn hope that I had suddenly turned into Ava Gardner.

It soon became clear that it was Mimi who had attracted the admiring glances. A photographer, who had come to take my picture for an article, spent only five minutes on me – the average time is usually 45 – before blurting out excitedly: 'I'd really like a photo session with your dog.'

Mimi enchanted another great woman of the 20th Century. Through my father's involvement in racing – he was chairman of the Horserace Totalisator Board – he became friendly with the Queen Mother. Twice a year she came to dinner at our house.

Eight months after Mimi's arrival, she was due for her summer visit. My mother had to submit a list of guests to the Queen Mother's residence, Clarence House, for her approval. But she omitted Mimi.

Guests, including Tom Stoppard and Peter Ustinov, arrived resplendent and grave in black tie, the women even more resplendent in evening dresses. At 8pm, a large black car drew up outside. Out stepped the Queen Mother.

She wore a long chiffon dress and appeared to be wearing the contents of King Solomon's mines. A huge ruby necklace nestled on her breast, matched by an exquisitely crafted pair of earrings – the gift of some long-dead potentate – while a diadem glittered in her hair.

Mimi, who had been told to stay in the kitchen, sneaked into the sitting room. At once, she decided to meet this human queen on equal terms.

Mimi leapt onto her lap, licked her martini glass, and began to chew her ruby earrings as if they were lumps of Pedigree Chum.

My father reacted with a cowardice of which he should have been heartily ashamed: 'That dog's nothing to do with me, Ma'am. It's Petronella's.' The Queen Mother merely giggled and stroked Mimi with her gloved hands. On subsequent visits, her first request, after a martini, was to see Mimi.

As well as sharing my highs, Mimi cushioned my lows. In 1997, when my father was diagnosed with cancer, her loving spirit provided the greatest succour. My father had chemotherapy in hospital during the week but was allowed home for the weekends, and Mimi made the decision to sleep at the end of his bed, where she could watch over him.

The doctors had fitted into his body something called a Hickman line that pumped medication into his chest even when he slept. My mother was worried that Mimi would damage the tube.

My father was mournful. His eyes welled up. He begged my mother to let Mimi stay with him. 'She'll think I've deserted her,' he said.

One morning he began to cry. 'What is it?' I asked. His reply was heartrending. 'I dreamed last night that I couldn't play with Mimi anymore.'

The day he died, Mimi started to choke. To have lost her too would have been unbearable. She survived and assuaged our grief. It was impossible to sit in a darkened room and weep because Mimi demanded my continuous attention.

Now I believe she did so deliberately, worried by my lethargic depression. I moved into a small house with a tiny garden until last March, when I got engaged to an Austrian and went to live in Vienna. My fiancé disliked dogs and, reluctantly, I left Mimi with my mother.

Last September, our vet, Bruce Fogle, rang me to say that Mimi had kidney disease. It was incurable but they would try to keep her alive by washing her kidneys with an intravenous drip.

I asked how long might she live. Eighteen months, he said, but it was unlikely. I put down the phone and gave way to tears. I cried during lunch, all afternoon and all through dinner.

The next day I cried through breakfast. My fiancé found this inexplicable. 'It's only a dog. It's 13 years old.' He told me I was making a spectacle of myself. and that I needed to pull myself together.

Three weeks later, I pulled myself together and

took a plane home, alone. It had taken a terminally ill dog to show me where my heart truly lay, and it was not in Vienna.

I had four months with Mimi. During her illness, she displayed an unflagging courage and patience – greater than most human beings are capable of.

The kidney washes lasted a whole day and her legs were punctured with needle marks and bleeding. But, to the astonishment of the veterinary staff, she never protested, never barked, but remained affectionate, gentle and stoic.

Gradually, her little body filled up with toxins and nausea destroyed her appetite. In early February, she was treated by a wonderful man called Grant. One grey, unforgiving morning, he called me and said we had to make a decision.

Mimi had no quality of life. She had fought so hard, but it was too much. It was time for her to be put to sleep. All of me cried out against it, but I knew I had to consent.

The night before, I slept with her beside me on the pillow. She was exhausted, but managed to lick my face. I felt a kind of anguished horror. In effect, I was her executioner. How do we know if a dog wants to die? The urge to live is so strong and she had loved life so much. But the alternative was watching her starve to death.

In the morning, I took her by taxi to the clinic. I think the cabbie spoke to me, but I can't remember. Mimi lay on my lap in her old, trusting way. I took her into Grant's office. I tried to be brave, but I lacked the courage of my dog. I couldn't speak.

When Grant asked if I wanted to be with her while she died, I could only shake my head feebly.

He took Mimi out of my arms. I kissed her nose and she gave me one last lick. It was like a benediction. She died showing a splendour of the soul, uncomplaining, resigned, loving.

When my father died, I stood over his coffin dry-eyed. But now I cried unceasingly at my loss. I drank, smoked 20 cigarettes a day and shut myself in the house.

As I type these words, the tears are starting again. But now I have Minnie. I had agonised over whether or not to buy another dog. Mimi, I convinced myself, was irreplaceable. She was unique – not like a dog at all, in fact – more like a human being. No puppy could compete with her memory. Moreover, this time, I was on my own.

But as the weeks passed, the house seemed horribly empty and quiet. There were no joyous barks when I opened the front door; there was no wriggling bundle to catch up in my arms; no playful companion to roll on the floor with, erasing the day's travails.

Men were no substitute. They failed to make me laugh and answered back.

I tried to contact the breeder from whom I had bought Mimi only to find she had died, so I trawled the internet until I found another Papillon breeder in Hampshire. She invited me down to look at her puppies.

I decided to go on a recce, but to wait until after Easter before making up my mind so I could be sure of my commitment. So it was that I met Minnie. Like Mimi, she was white with brown markings, but her ears were already so big she resembled a canine Mr Spock. Her little pink tongue licked my face until she was covered in foundation. Suddenly, my doubts evaporated.

Driven by some inexorable force, I wrote out a cheque and took her with me.

Minnie has already taught me that all dogs are unique. She is even naughtier than Mimi and more excitable. She has the agility of a gazelle and leaps from chair to table.

Her intelligence is astonishing. She takes a great interest in my work – though sometimes I wish she didn't. After discovering what a marvellous toy a computer is, she now uses it herself.

As I write, Minnie is chewing through my computer

lead, having contracted ennui after biting off the heels of a pair of Manolos. The day before, she had destroyed a white Chanel skirt.

Last week, she deleted most of an article over which I had taken great trouble. A few days later, I tried to book a flight to Budapest, foolishly leaving my computer with Minnie while I went downstairs to collect my post.

When I received the easyJet confirmation email, I was astonished to find that I was booked on a plane to Bucharest.

I examined the computer, which had Minnie's footprints all over it. I rang easyJet and explained it was all a mistake. It is no easy task to convince airline staff that your dog has booked a ticket to the wrong destination.

I was politely called a liar. Resignedly, I shut Minnie in the kitchen and forked out another £100.

She wakes me every morning at 6.30, biting my nose. I take her outside to pee. She doesn't. Instead she does laps like a canine Lewis Hamilton. I bring her back inside and she pees immediately.

She booby-traps the house with squeaking toys, removes my lipstick from my handbag and paints the floor Jungle Red. But I can't do without her. I am growing used to rising at puppy-crow and my initial exhaustion has been replaced with a new energy.

I can't say I enjoy scrubbing the carpet after she has peed indoors for the fourth time, but on the other hand it has done wonders for my bingo wings. I am fitter and healthier. I hardly drink because Minnie is a far more effective stimulant. I have cut down on my smoking because I don't want her to breathe in the fumes. I jog every day. Minnie insists on an hour's 'walk', spent running as fast as possible. After three weeks, I have dropped a dress size.

I am also happier. Minnie, with her slender frame, elfin eyes and natural elegance, promises to be the Audrey Hepburn of dogs. As Rex Harrison sings in *My Fair Lady*: 'I've grown accustomed to her face. She almost makes the day begin.'

I've grown accustomed to her eating my earplugs to ensure I have no lie-in on Sundays; to her sharing my breakfast (she gets the toast soldiers and I get boiled egg on my face); to watching her make my home her own. I've grown accustomed to her sleeping next to me at night, her funny face exhibiting both a vulnerability and a desire to protect. She makes me feel safe and warm, like the promised kiss of spring-time. It is Easter and the sun is shining. Minnie is now using my little finger as a chew toy. I am lucky. I have been twice-blessed.

Buster was a rescued cross-breed who became my best
friend and remained my loyal companion for 15 years.

CHAPTER 4

BUSTER
A shameless display of emotion

ROY HATTERSLEY

Truly, it's the little things that I miss the most. The tinkle of his medallion when he ran to greet me. The smell of wet dog as I dried him after a rainy walk. The rattle of his bowl against the kitchen flagstones during the 30 seconds he took to eat his breakfast.

Then there was his sudden appearance in my bathroom when my shower took longer than he thought reasonable, and the look of deep resentment if he was sprinkled with water as I reached for a towel.

I even miss the old causes of annoyance. These days, I can load the dishwasher without fighting a losing battle to stop Buster licking the plates, and I can leave the morning's letters on the doormat without them being perforated by Buster's teeth.

Now, I long to be inconvenienced again: to be forced by Buster's persistence to go out in the freezing Peak District rain; to be woken in the middle of the night by his snoring; and to go through the complicated ritual of fastening on his safety harness in preparation for a journey.

In the car he would fall asleep and, if we were driving from London to Derbyshire, wake up with a whoop when we turned the corner into our village.

After he had inspected the house, room by room, he would sit on the first landing, staring out of the window and grumbling at the ramblers who changed into hiking boots sitting on our wall.

We played a game on the landing. Buster had to guess which of my hands, on the stair below him, held the biscuit. He pawed at my fingers without ever hurting me, and always won the biscuit in the end. It is one of the little things that it hurts me to remember.

Most dog owners regard their dogs as special, so I do no more than describe the qualities that I found irresistible. Thanks to his energy, he imposed himself on all the lives around him. Doors banged open as he marched into the room. A morning rarely passed without him becoming entangled in the wires that connected my laptop to the world.

I could never lay a fire in the drawing room without him helping me by examining every log. When he saw bags in the hall, he sat among them — like a brindle suitcase — to make it clear that he was travelling, too.

A guest who sat on what he regarded as *his* sofa

often found that Buster hurled himself into the next seat and leaned hard against his new friend, head on shoulder.

He was not so well-disposed towards cats, rabbits and domesticated rodents. But he liked people.

He became a favourite at book festivals. Literary old ladies travelled across country with 'treats', which he never refused.

While I was speaking, he only barked during the applause or when, by putting my hand in my pocket, I gave the impression that I was about to produce a treat myself.

I have spent long hours during the past ten weeks thinking what it was — apart from the thrall in which I'm held by dogs in general — that bound me to Buster. I enjoyed the knowledge that he was dependent on me, and I admired his apparent belief that I was dependent on him.

His appeal was increased by what is best described (despite the reputation he acquired after he killed one of the Queen's unfortunate geese in St James's Park and I was fined for contravening Royal Parks' regulations) as an aggressively affectionate nature.

But, most important of all, he radiated hope. Whenever I opened the pantry door, he appeared behind me — assuming that I was getting something for him.

I would call him a born optimist, but I never made the anthropomorphic mistake of thinking of him as a little man in a fur coat, and dogs are not capable of thinking about the future.

He was never fed at table and he slept in his own bed. Treating him like a dog was a mark of respect. Being a dog was enough. I asked for nothing more.

For 15 years, I watched him grow up, grow wise and grow old. His vet predicted he would be happy to the end but that, one day, he would just be too tired to carry on. 'When it happens, he will let you know,' I was told.

And so he did. Every step of his brief morning walk was a struggle. Breakfast was eaten with slow determination. Then he lay down with no intention of ever getting up again.

The final decision had to be based on what was best for Buster. So the temptation to put off the fatal decision was resisted. After a moment of agonising indecision, I made the fatal phone call. The vet arrived within the hour.

Buster died eating a piece of blue cheese – the much desired but forbidden food which he usually only tasted when it was wrapped around a pill.

I do not pretend that my grief was unique. Many families, I know, have been devastated by the death of a dog. I merely state, as a matter of fact, that nothing has ever caused me as much pain as Buster's death.

Nor have I ever behaved with such a shameless display of emotion. I sat in the first-floor room in which I work, watching my neighbours go about their lives, amazed and furious that they were behaving as if it was a normal day. Stop all the clocks. Buster was dead. He left a permanent legacy. Do not underestimate what a dog can do.

I never contemplated teaching him to sit-up-and-beg, shake hands or play dead. And Buster certainly never condescended to carry a rolled-up newspaper in his mouth or retrieve balls.

But he *did* – perhaps it is a minor achievement – change my life.

Some of the ways I can describe: I gave up red meat because I could not bear the thought of eating anything that was Buster-shape.

But there is more to his abiding influence than that. His real legacy is the memory of the pleasure he provided.

Birth and upbringing – and an almost-labrador called Dinah – made me a dog person. Buster ended the years of deprivation and made a return to dogless life unthinkable.

Of course, there were times when it seemed that even the thought of another rescued cross-breed in the kitchen was a betrayal. But after ten weeks, the search began. A new dog would not be a replacement, as Buster was irreplaceable. His successor would be a dog in his own right. But he would be a reassertion of all that Buster stood for; the incalculable blessing of possessing a dog.

I told myself that he – it was always going to be a he – would be a tribute to the memory of Buster.

But I also felt guilty. Thinking about another dog waiting on the landing to welcome me home seemed like a betrayal.

So I began half-heartedly to look at dog sanctuary websites. There was a mad moment when I thought of adopting Harley, who was advertised as a loving chocolate brown half-Labrador.

He was all those things. But the other half was Bull Mastiff. Harley – six months and still growing – weighed six stone. When, as a gesture of affection, he leant on me, I fell over.

I also looked at Nelson – lacking one eye, as his name implied – when a Kong dog toy rolled towards me from the next kennel. A Kong is a lump of rubber in the shape of a deformed pear, which delights dogs by bouncing at unpredictable angles.

The Kong was delighting the kennel's occupant, Jakie – not the mongrel I was looking for – but a purebred English Bull Terrier pup. I threw back the

Kong. Jakie rolled it out again, and so we began a rally of Wimbledon proportions.

I discovered he had been found wandering the streets of Bolsover in Derbyshire just before Christmas. He was, I was told, desperate for company.

In the rescue centre, surrounded by dogs but alone in his pen, the Kong was his only consolation.

I decided there and then that Jakie would have another friend, and that it would be me.

It was not love at first sight. That happens only once in a dog-owning lifetime. So I took Jakie home and waited for him to grow on me.

The process was accelerated by two misfortunes that turned out to be blessings. First, Jakie was ill, not seriously but enough to stimulate the protective feeling that is close to devotion. Then a London neighbour complained to the local freeholders' association that I had bought 'an attack dog'. I can only guess what gave her that idea. She had seen him only once and had not mentioned her concerns to me. I assume she judged him by his appearance – a test that few of us would pass. To most people, he looks more comic than threatening. It is part of his character.

Jakie has the huge head and snout common to his breed and a typically thin coat through which there's a glow of pink skin. His brow is creased in permanent wrinkles, as if he is worried or thinking deeply, and his big lips give the impression that he has smudged his lipstick.

On a clear day, light shines through his permanently erect ears. Perhaps his almond-shaped eyes were thought to be a sign of Oriental menace.

My neighbour's complaint came to nothing. But for the couple of days before the estate manager recognised a breed known to the Victorians as 'the gentleman's friend', I relived all the irrational fears I had experienced when my previous dog, Buster, killed the Queen's goose in St James's Park. Then, though nothing worse happened than a £200 fine, I was sure – out of pure neurosis – that the police would come to take Buster away. I was ready to barricade the doors, hire the most expensive QC in the land and flee (with Buster) to Ireland.

After this new complaint, I felt the same about saving Jakie. No one would take him from me. He was mine and I was his, till death us do part. Perhaps he had arranged the complaint as a bonding exercise.

So the serious business of training began. He was, miraculously, already house-trained, but he had no idea how to deal with stairs, and I doubt if he had ever been taken for a walk in his young life. Teaching him to proceed in one consistent direction was the first lesson. He learned quickly.

Now we cover long distances at light infantry pace each morning in the hills of the Peak District, where we spend most of our time, and in London's royal parks, where we avoid birds.

After an intensive six weeks, on the word of command he will sit, go flat and stay – more often than not.

He is friendly towards sheep, which is only to be expected since he looks like one.

I have not become the love of his life. He remains true to Kong – well, the series of Kongs with which he has been indulged.

Offer him a biscuit or a Kong, and he will take Kong every time, chewing it while it sticks out of the side of his mouth like a grotesque cigar and pursuing it across the carpet after its frequent escapes from his teeth.

The second place in his hierarchy of esteem almost certainly belongs to the hard blue rubber balls which – were I prepared to throw them for him – he would chase round the Derbyshire garden all day.

His fielding is one-day Test-match standard and his sliding stop smears his white fur with the sort of mud and grass stains that disfigure cricketers' flannels. Unfortunately, Jakie does not wash.

He dribbles his ball with the finesse of a Premier League footballer. One way and another, he's ruining the grass. But it's a small price to pay for a close association with an all-round sports star.

Chasing balls, Jakie shows no concern for life or limb. He crashes into walls, falls down steps, bounces off apple trees and disappears into rose bushes with no thought other than retrieving and responding to the order 'drop'. Not bad, I tell myself, for a dog who hates going out in the rain and looks nervously over his shoulder if he hears a loud noise.

Jakie is very different from Buster. And that is important for both of us. I loved Buster because he could lay claim to being the hardest dog in the world. I am growing to love Jakie because he could not. But the bonds of affection are unbreakable.

In the early days of doubt, when I wondered if another dog was a mistake, I told myself that at least he had found a good home and that he was lucky to have me. Now I think I am lucky to have him.

I didn't choose Jakie; he chose me, which was just as well. Had he not imposed himself upon me, I might be missing the joy of a friendship that has enhanced human existence since cavemen tamed wolves. Now, with me on one end of a lead and a dog on the other, I walk with a spring in my step again.

DOG STORY

Lady Annabel never met a character like Copper. For 15 years, they enjoyed a rollercoaster life together which, although not always smooth, was certainly never boring.

CHAPTER 5

COPPER
The greatest character I ever knew

ANNABEL GOLDSMITH

With five loving children (sadly I lost my eldest
son Rupert in 1986), I consider myself one of the
luckiest women alive. Although I have a large circle
of friends, it is my family, including my sister and
brother, who matter most to me in the world. My
life has been packed with incident, but it has never
been lonely, as my family have always been there to
share times of great happiness and to sustain me
when things have not been so easy.

Dogs have played a fundamental role in my life.
Having grown up with the little stone monuments
to dogs and horses in the Peace Garden, the family
pet burial ground at Mount Stewart, and those
that are scattered around the grounds at Wynyard
Park, including the grave of that great racehorse
Hambletonian, I now have similar memorials at
Ormeley commemorating my love for my own dogs
that are buried in the garden there. A dog's love is
unconditional and its companionship unsurpassed,
and I cannot imagine my house without a great

number of them. And whenever I have been away, my return is made happy by the welcome they give me. Robin, my second son, has inherited my love of dogs, and a few years ago when my two Norfolk Terriers Barney and Bee had four puppies, Robin rang me to tell me to look in the births column in *The Times*. 'To Mr and Mrs Arthur Barnes,' it said, 'three beautiful daughters and a bouncing baby boy', followed by the full address at Ormeley. Eager to capitalise on the prospect of a new, really substantial quadruplet account, a well-known shop specialising in baby goods sent me a complimentary hamper-sized supply of disposable nappies the following day.

Currently, there are four dogs in residence: Lily and Daisy, named by my grandson Sulaiman, are officially Grand Basset Griffon Vendeens, which translated means large white shaggy French gundogs. This breed is famous for its friendliness and good nature, and Lily and Daisy are no exception. They have been joined by two mongrels – Scruff, an amiable Alsatian cross, and Poppy, a Lurcher Collie cross. Sadly, the Norfolks, having lived to a ripe old age, have died and are buried in my dog cemetery, each with a little inscribed headstone.

It is unthinkable for Lily to greet my visitors without bringing them a present: sometimes one of my shoes or, more frequently, a pair of my pants that she has retrieved on her regular forays into the laundry basket.

Shortly after Lily's arrival, I paid a visit to Rigby & Peller, the lingerie shop in Knightsbridge, to buy some new and rather large bras. At about the same time, the trial of Paul Burrell, Princess Diana's butler, was approaching and I had happily agreed to talk to Paul's lawyer, Andrew Shaw, later that same day. Although I had very little to contribute, I liked Paul and knew that Diana had been very fond of him.

Mr Shaw arrived at the house soon after my return from the shop. He and I were deep in earnest conversation when I noticed the door opening very slowly and Lily coming into the study, slowly wagging her tail. She had something dangling from her mouth which she laid lovingly at Mr Shaw's feet. I saw to my horror that it was one of my vast new bras, and a purple one to boot. She laid it out in all its tremendous colourful glory, resembling, as my brother Alastair would say when he spotted it hanging on the washing line, 'a parachute masquerading as a brassiere.'

Crimson in the face, I tried to reclaim it but Lily playfully snatched it back, running around the room, pulling it by the straps and showing it off from every angle. When I finally managed to wrench it away from her, I found it almost impossible to bring myself to look at Mr Shaw's bemused face.

It was, however, my mongrel Copper, whose intelligence far outshone any lack of blue blood breeding, who was the greatest of all the characters I ever knew. I found Copper in a butcher's shop, not on the meat counter with the legs of lamb and joints of beef, but in a pen surrounded by his siblings, a multitude of puppies of indeterminate colours and dubious origins. I am not sure why I chose him above the others, but I know he was copper-coloured and floppy, and that when I first saw him he was lying on his back. Against my better judgement, I had promised Jemima I would give her a puppy if she passed her Common Entrance Exam, and, whilst I had no way of knowing what he would grow up to look like or how large he would become, I took him home with me and placed him in Jemima's arms. She fell in love with him on the spot and, as he was so small, she would sing him to sleep in her hands. Within a few months, he had left puppyhood behind and, as he grew bigger, the distinctive curly tail of a mongrel appeared. With his dignified beard, he had become a young dog.

Rather like the hero of one of the children's favourite films, *The Belstone Fox*, Copper became a living legend in the neighbourhood and notorious for his exploits.

COPPER

By his first birthday, he had learned the bus route to Richmond and Kingston. However hard I tried to keep him in, he would escape to Richmond Park to chase joggers, rabbits and anything that moved. And when he felt a bit peckish, he would take the number 65 bus into town, stopping off at the Dysart Arms on the way, where, if he was lucky, he would be given lunch. Sometimes he would be accompanied on the bus by my black cat Jessie. Copper was equally partial to a pub on the outskirts of Kingston, and more than once he was spotted by one of my friends sitting up at the bar, lunching with an unidentified elderly man. He was also seen in pubs as far away as Surbiton and East Molesey.

At dusk, he would make his way home by trotting through the woods and waiting by the pedestrian crossing on the busy Petersham Road. There, he would pause, hold up his paw quite pathetically until the cars stopped, and cross over with a heavy limp. On reaching the other side, he would scamper off home. I did not want to lock him up as that would have made him miserable, and as years went on I became less worried about his expeditions because he always made it home in time for bed.

His amorous adventures were in a class of their own. He could smell a bitch on heat from miles away and must have fathered more puppies than the highest paid stud dog. I will never forget one incredibly cold and beautiful winter's day, when the snow lay thick on the ground and the pond on Ham Common was frozen over. The branches were heavy with snow and dripping with icicles. Zac and Jemima put on their skates and joined the other children on the pond. Copper was amusing himself by pulling Benjamin around the pond on his sledge until, getting bored with the game, he wandered off.

As I was watching this idyllic scene, I was joined by a very snobbish lady who lived nearby and whom I vaguely knew. She was the proud owner of a very well-bred white standard Poodle bitch, a regular entrant at Cruft's dog show. She told me in a breathy whisper that in a few days' time she would be sending her beloved pet to stud, that she had a queue of people wanting puppies and that each puppy would be worth hundreds of pounds. She even mentioned the intended pedigree sire's name, which was particularly long and pretentious. As she droned on, I happened to glance behind me and saw that Copper had mounted the white Poodle behind the sycamore tree. Terrified at the possibility of the owner turning around to see what was happening, I engaged her in frantic conversation, pointing out interesting sights on the pond in front of us and feigning disapproval at some of the younger skaters' bad language. Furtively peeping over my shoulder, I could see that by now the two dogs were 'tied' back to back; a sure sign of a successful mating. I managed to keep the white Poodle owner distracted, wondering how on earth she had failed to notice that her dog was already on heat, until eventually the dogs separated and the Poodle raced around clearly delighted with herself and her new conquest. Copper emerged looking bedraggled and exhausted. Luckily, the white Poodle owner's family moved away shortly afterwards, but I have often imagined the look on the owner's face when the Poodle's puppies were born.

Copper soon moved on to my daughter India Jane's mongrel, Dushka. Dushka did not fancy him as much as the white Poodle, but she had little choice in the matter as she managed to get herself stuck in the cat flap while trying to escape Copper's attentions. He moved in fast. Whether or not he was the father of the two puppies she produced remained a matter for conjecture, but I think a Dachshund must have got there first. Neither of the puppies looked much like Copper, but he became very attached to the one we called the Platypus, who was black with a very long body and short little legs. The Platypus was keen to accompany Copper on his many adventures but, unable to keep up the pace because his legs were too short, he would soon be dumped by Copper on the nearest doorstep. I was getting at least two calls a day from people asking me to come and collect him.

Once, Copper took the Platypus to Brighton for the day. Trotting down the side of the A3, they were given a lift by a charming lady who owned an antique shop in Brighton. The woman, upon seeing this extraordinary duo by the side of the road, had thoughtfully called my number, which was printed on the inside of Copper's collar. After that, the Platypus went to a good home in Herefordshire and Copper resumed his solo roaming.

But his freedom was soon curtailed when he was accused of biting a jogger outside the house. A summons arrived and I prepared for Copper's day in court. I found an excellent barrister and took masses of photographs of the driveway where the offence allegedly took place. Among the pictures I enclosed for the magistrates as evidence of Copper's innocence was one of him looking rather rakish, dressed as a member of the Mafia, wearing a hat and dark glasses, a cigar clamped between his teeth. In court, I watched as the magistrates tried to suppress their grins upon seeing the picture. Eventually, helped by many character witnesses, including dear Jilly Cooper (who had loved him ever since she judged a charity dog show in my garden and found him to be the naughtiest dog in the show), Copper's life was spared. The penalty, however, was that he was no longer permitted to roam free.

As the years went by, Copper settled into contented old age, spending much time wrapped in his duvet asleep. But one winter afternoon, I took Copper and my Norfolk Terriers for a walk on the golf course adjoining Ormeley – a walk I often took by arrangement with the golf course after all the golfers had finished playing. Half an hour later, I returned with the Terriers and left the gate open as usual to allow Copper to return in his own time. Ten minutes elapsed and still he had not returned, so I took a torch to go and look for him. Such a search would have been pointless when Copper was younger as he was a law unto himself, a famously free spirit who came and went at will. I searched for him in silence now, as he was stone deaf. It had been his only concession to old age. Thinking he might have found a hole in the fence and escaped, I got in my car and drove up Ham Gate Avenue, parallel to the golf course. I thought I had spotted a dead fox lying by the side of the road and was about to avert my eyes when a terrible thought struck me. There was my beloved dog lying in a pool of blood. He must momentarily have lost his instinctive road sense and been hit by a car. No words can adequately describe the feelings I experienced when I found my dog run over. I tried to accept the awful violent end of a life as I gathered up his frail body and wrapped it in a towel. At fifteen, Copper was an old dog and I had known he would not live more than another two years, but I had hoped he would die peacefully on the sofa in my bedroom, wrapped in his Colefax and Fowler duvet.

We buried him in the wild garden at Ormeley and my brother Alastair composed the following words for the tombstone:

Copper
1983 – 1998
In loving memory of the professor,
A true gentleman of the road.
Tragically removed from our midst
By cruel fate but never to be forgotten.
As the poet has said,
'They are not dead who live in hearts
They leave behind.'

DOG STORY

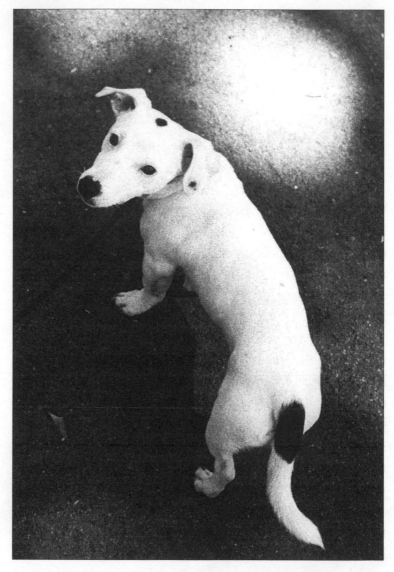

Love is Mr Pig. Mr Pig, born August 1992, passed away at
five o'clock on Friday 10th September 2004.

CHAPTER 6

MR PIG
Not a dog at all

PHILIP TREACY

Walking with me into work along the Battersea Road early in July in 2004, Mr Pig just stopped, sat down and would not walk any further. He wouldn't move. I told him we had to get into work but he just sat there. I said 'Mr Pig what's wrong?' Because he looked fine, I didn't understand that he was in pain.

I was on my way to meet some customers, so I took him back home to my partner Stefan and asked him to take him to a vet in Battersea. When Stefan returned he told me: 'We have to go to the animal hospital in Hatfield now.' I asked why and he just said: 'Because Mr Pig has a tumour.' With no time to think, we drove straight there whilst I took in the implications of Mr Pig having cancer.

The Queen Mother Hospital for Animals run by the Royal Veterinary College is an incredible place. It was Mr Pig's first time in a kennelish environment and he was not used to it. Mr Pig never went to a kennels; he wouldn't have understood a kennel. A kennel would have been the most horrific experience.

But he had no choice as he had to have an operation to cut out the tumour straightaway. Leaving him at the hospital, and not wanting to upset him, we decided it best not to see him for a couple of days. Everyone thinks that their dog is the most incredible dog but he was very, very special and I was very, very concerned about him. But I thought: 'Let's cut the tumour out and keep going.' The tumour was in his stomach and the vets explained what the little white marks were on the x-ray.

When we went back to see him after the operation there was a kennel area and they had obviously picked up on Mr Pig's sense of self-worth as his kennel had a crown on it. There were 20 dogs and they were all ill, but he was the important one. Mr Pig was absolutely delighted to see us and thrilled to be going home. He was celebrating on my lap all the way home.

I hoped he would be back to normal but he had a great big scar down his stomach which was a bit worrying to look at. But he seemed great so we went on holiday to Oxfordshire where someone had lent us a house. Life was back to normal and I just presumed it would all be fine.

But then a few months later, I was walking him around Chester Square and he stopped again and I had to carry him home. I went home and told Stefan that he had stopped again, so Stefan said: 'Well, you know you have to understand that it might come back' – which I had not thought of.

I remember it being a defining moment.

But he improved again and we walked to work in the morning until one particular day on 5th September. We had gone to work and we went to the park at lunchtime. I bought him sliced ham and we sat in the park together and I had other delicious things and we hung out there for a few hours. I came back to my studio and he was standing in the kitchen – just standing there – and I sensed that we had better go back to the vet to get things checked.

By then it was around four o'clock when we got to the vet, and we saw a very nice man. He could see the panic in our faces. He took Mr Pig away for a few minutes and did some x-rays. At that stage, I really thought Mr Pig was just having a check up. The vet came out and said: 'You know, the problem is that the cancer has returned and we can cut it all out again but it will return.' The vet could see that we did not know what to do. He was a very sympathetic man and he explained that they could do the operation again. But he said that did not mean that it would not return again. I said: 'What would you suggest? What is the best for him?' He said: 'Honestly, I would suggest that you would let him go now.'

And I remember thinking 'now'.

He told us we could take him with us that evening but that it would be quite hard for us to be with him knowing that we were coming back in a few days to do the same thing, and that sounded horrific.

He said: 'Look, you can take him home for the weekend and you know in a few days time to bring him back, but it could be worse knowing.' I said: 'Give me a few minutes to think about it.' I talked to Stefan and we were both crying and wondering what to do. Mr Pig seemed fine. He did not seem like he was going to pop off any moment and he was not that old. Everyone knew that he had had a tumour cut out and was not that great, but I did not think that it was the end of the world. I thought they had got it and cut it out.

I knew we had to do what was best for Mr Pig and we decided to take the vet's advice. Animals are so helpless because they can't tell you what is wrong or how they are feeling. The thought of him just going quickly like that was very difficult, and it was the worst thing I have ever had to do.

The vet went back inside to get Mr Pig and he placed him in my arms. He was a little bit drowsy and he had a needle attached to him. He just looked at me – we looked at each other.

He was so sweet lying against my heart and, like that, he just gently went. As I was holding him, I could tell he knew I loved him and I knew he loved me, and it was as simple as it was. As he was dying, he lifted his head and laid it on my shoulder.

He had been sedated so he did not know what was happening and he was not upset. It would have been awful to have been at home with him and in pain.

That day, Friday 10th September, at 5pm was the most painful moment of my life.

I told the vet we would be having a funeral and I would have to organise it. He volunteered to keep Mr Pig until then and said that the hospital would look after him.

So Stefan and I left without Mr Pig.

We walked home absolutely devastated. We had thought that we were going for a routine check up and it was shocking to be leaving without him. We just cried all the way home. Mr Pig had been my family for 12 years and now suddenly he was gone.

I telephoned the studio to let everybody know what had happened so they could tell people.

The news quickly spread around our friends and family, and that night we had a wake; a proper Irish wake. I am Irish and we like to deal with loss in a therapeutic way, talking about it and telling everyone rather than pretending that it did not happen at all – which is worse. As news got around that he was dead, people were phoning and everybody who was near came over to celebrate Mr Pig's life. Gradually the house filled up and that night there was a lot of conversation about Mr Pig. He was my relative and everyone knew how important he was to us.

For me, it was absolutely horrific because he had been through everything I had been through in the last ten years; every high, every low.

After the wake, I planned Mr Pig's funeral. It sounds ridiculous, but only animal people will understand how he deserved it. We had a little coffin made with his name on it and it was sent to the vet.

On the following Friday, the undertaker went to the vet's, collected the coffin and brought Mr Pig to Elizabeth Street. I could not bear to see Mr Pig again so I asked a friend to wrap him in one of my t-shirts. We had the funeral in Sylvia Pugh's garden. Sylvia lives above my shop and has a beautiful garden in the back. We showed a video some friends of mine had made, featuring the highlights of Mr Pig's career. Grace Jones sang a song that she had written specially, which included words I will never forget: 'Mr Pig, he really did not know that he was a dog at all.'

It sounds completely ludicrous at a time when people were being buried alive in earthquakes but I saw Mr Pig as my friend, not my dog, and he had his own way of conveying his love and devotion to me, and it was the least I could do after he died.

I tried to get him an obituary in *The Times* and they thought that it was hilarious and did not take the request very seriously. But we did get a call from the *Sunday Times* 'Style' magazine, which carried his obituary on 10th October 2004.

For days afterwards, I did not really know what to do and work was the last thing on my mind. It was not really of interest. Mr Pig was my everything, he was like my child. Because of the hopelessness of it and the sort of love he had for me, it affected everyone around me. I remember a friend of mine in Los Angeles calling from work and just starting to cry. There was no finality to it, and I didn't particularly want there to be.

Mr Pig arrived in my life in August 1992. At the time I was kind of homeless, living in my studio in Ebury Street. A friend called Yvette Jelfs came in one day with two tiny Jack Russell puppies.

They were two tiny little pups and she wanted to give me one of them.

I hadn't asked for a dog and I wasn't looking for one, she just thought I should have one.

She pointed to one of them and said: 'This is Linda,' and I said: 'But who's that one?'

He was this tiny little one with a black spot in the middle of his forehead. After not wanting a dog and not being vaguely interested in having a dog, I said: 'I like him' and had no hesitation whatsoever in saying: 'Yes.'

And that's how Mr Pig arrived in my life. I was just thrilled as I'd never had my own dog but I had grown up with Jack Russells in Ireland so I knew their temperament well.

Jack Russell puppies are very sweet little things, and he looked a little bit like a piglet. I was transfixed by his black spot, which was perfectly in the centre of his forehead. He was too small to walk so I used to carry him around. People would ask: 'What is it?' I would say to some: 'A piglet' and then to others I would say: 'It's a Dalmatian' because he had one spot, and they would say: 'Really?' He was really very white and pale pink as a puppy, so I just thought Mr Pig was a good name because it was very snappy and direct.

I had named him almost immediately as I just thought he was a real 'Mr', this one. He had a lot of attitude. I was sleeping on the floor in my studio on Ebury Street. He had to sleep in the palm of my hand because he was so tiny. So I put my hand over him at night and held it there all night long so that I wouldn't roll over and squash him. We were friends immediately and he would hang out all day long – as long as I was here with him. There's something very touching about an animal that is dependent on you for everything, and I realised that I had a special connection to Mr Pig straightaway – the love that those eyes can convey. We became very close very quickly.

I was 26 years old and Mr Pig was 6 weeks old.

Many people thought it was a mistake having a dog because I didn't really have anywhere to live and they believed that Mr Pig would not be looked after properly. I told them we would work around it somehow.

Soon afterwards, I had to move out of the studio in Ebury Street and into the Irish Club on Eaton Square. It was a beautiful building but the accommodation was beyond Victorian and dogs weren't allowed in. But Mr Pig was living there with me and I would have to sneak him in and out in a rucksack. I used to put the rucksack on my back with him in it.

The longer I stayed at the Irish Club, the more I got to know the men on the door and the more they wanted to chat to me in the mornings, and the more complicated it became. I knew I had a dog, in a bag, on my back and it was a little tricky at times as I knew if he started barking he would blow our cover and we would be sleeping rough. He never did, but I could always feel him moving. This went on for a couple of months, which felt like a couple of years because of the pressure of getting him in and out. Not getting him to make any noise was getting more and more difficult. In the end, I had to move into my shop on Elizabeth Street.

Right from the start, Mr Pig went everywhere with me and he got used to going all around town. He loved to pretend that he was on his own and he would always walk about 30 feet ahead. I always found it funny because, although he would pretend that he was out on his own, he would always check that I was coming behind him. In all our years together, Mr Pig was never on a lead and, once, when he was unfortunate enough to get run over by a car, he was clever enough to avoid the four wheels as the car drove over him. That was the closest we came to a disaster.

He was the first dog that was allowed into the BBC building in Portland Place. I went there to do a programme with Esther Rantzen and when I got there the doorman said: 'Dogs aren't allowed.' I said: 'Look, I'm really sorry but unfortunately if he can't come in, then I can't come in either. He is with me.' So they had to get special dispensation for Mr

Pig to be let into the building. During the interview with Esther Rantzen, I looked down and saw Mr Pig chewing her scarf and thought: 'If she looks down now we are all in trouble.'

There were other difficult experiences and I remember being invited to Elton John's house in Windsor. I was in two minds whether or not to take Mr Pig but I thought: 'Elton is a dog lover and it's not the end of the world, so we will work something out.' But when we got there, Mr Pig wouldn't stop barking in front of everybody. There were about 25 guests and he was sitting at the table next to me. Elton came around to say 'hello' but every time he went near him, Mr Pig would go: 'Grrrrrrrrrr.'

I put him down on the floor and said: 'Mr Pig, relax. This is slightly mortifying.'

Then Elton went to find some dog biscuits but Mr Pig was having none of it, and every time Elton John came near he went: 'Grrrrrrrrrr.' I always remember Guy Ritchie, who was at the party, saying: 'I can't believe you brought a dog to the party,' and I said: 'Well, neither can I.' Guy thought it was hilarious but I didn't think it was hilarious at the time, it was just a nightmare.

Over the years, as I became better known and as the company grew, I was always being photographed and I just never felt comfortable about it. So I began being photographed with Mr Pig. I thought this cute, gorgeous little dog will distract from the fact that I personally do not feel so cute and gorgeous myself and that people would look at the dog instead. So we became a double act and Mr Pig would pose in a certain way. When a photographer pitched up, Mr Pig would just jump on my lap. If the photographer did not want him, he was not exactly thrilled.

Mr Pig was always his own person, and he did not like everybody. When he took a dislike to somebody, I respected it because I'm not crazy about everyone myself.

People would always say: 'Oh dogs love me' but sometimes he didn't, and he would snap at and bite them. I understood why he did it as, invariably, they had pushed him too much and treated him like a little idiotic thing.

That is not what he was and he appreciated being treated like an equal and not like an idiot. He behaved in a very human way. In fact, he enjoyed conversations and would sit there and listen and sometimes, it seemed, almost partake in the conversation. In a meeting, he would sit at the table and just look at everybody, just sitting there like a child taking in all the conversations. We would sit and stare adoringly at each other and he gave the impression that he was in the meeting as well. He would just look across the table from person to person.

He behaved liked he owned me, not the reverse, and I was his property in a sense.

Mr Pig became a fixture at my shop in Elizabeth Street, and he got used to sitting in the front window. He would sit silently in the corner, looking out onto the street all day long. He would sit so silently and so still that he looked like a stuffed dog.

When people came up to the window to look at the hats, they would hear: 'Grrrrrrrrrrr' as Mr Pig turned his face to the side and showed them his teeth. That was what he was like. I would say to him: 'Mr Pig, we are trying to sell a hat here', but he behaved as if the shop was his territory.

He really did react to the customers as if he didn't want them to be there. He was very wary of some people.

But some people he liked. We would get a customer who wouldn't like dogs and they would say: 'Oh no, I don't like dogs' and would leave immediately. I would be offended for him, thinking: 'Well, excuse me, but he is more important than you as far as I am concerned', but I couldn't say that. When that happened, Mr Pig would have to go downstairs and he wasn't crazy about that idea.

It was just a random thing and some people he

really liked and some people he absolutely loved.

When he really liked someone, he would jump up on them and sit on their knee. But if they had heard of his reputation, they would be frozen to the spot.

They all thought Mr Pig was a cute little thing but he wasn't always such a cute little thing and some-time he just snapped. He bit people on the nose and sometimes there was blood and everything. So I had to be careful, I went upstairs in the shop one day and he was swinging from this woman's hand because she had pushed it just a little too far.

I remember Grace Jones coming to my flat one day when I wasn't there. She was watching television with Stefan. Suddenly Mr Pig just jumped up on her knee and started watching her. She was thinking: 'Is this really happening, is he going to bite me?' He did not like being treated like a little cute dog because he did not feel like a little cute dog. He felt like a person.

Of all my friends, Grace Jones was perhaps closest to Mr Pig. He had a small, special, squeaky red toy Wellington boot, but he played with it only when Grace visited. Mr Pig produced so much noise on that squeaky toy it was like he was playing a musical instrument. Grace would rap along with it, know-ing he was playing the boot for her. As he played melodically, she would applaud and say to him: 'Mr

Pig, you and I have got to record an album! You've got such talent.'

Once I made a hat for Prince Charles' wife, the Duchess of Cornwall. I was late getting to the shop and I hadn't met her before. As I was walking down the stairs, I could see Mr Pig kneeling. I couldn't see her but I could see him staring adoringly at someone, and I thought: 'This is interesting, Mr Pig.' The Duchess is a Jack Russell lover and she knew how to treat Mr Pig. From then on, Mr Pig used to receive his own Christmas cards from the Prince and the Duchess.

I once took him to the cinema and it didn't work out. It was to a late night showing of *Reservoir Dogs*. I didn't want to leave him at home so I put him in my jumper and took him with me. When the lights went down in the cinema, he started howling and barking and, as it was called *Reservoir Dogs*, everyone was looking for a dog on the screen. But I just had to leave.

When I went abroad, there were times I had to leave him with Isabella Blow and he would go off for weekends in the country on his own. Isabella knew he was my prized possession, so Mr Pig slept in a baby's crib in her room.

He also liked beautiful girls and so there were lots of girls around the studio. He preferred women to men and would go off at the weekend with Amelia Couttisson, who was my assistant at the time. She used to refer to him as her 'little white prince.' He was so clever and people quite liked having him but they knew he was difficult and that there would be a breaking-in period.

I always missed him when he went away, but I felt comfortable with anybody I left him with. I knew he would be having a great time. He was on holiday and it wasn't a sad sort of situation – well, most of the time anyway.

He used to stay with a friend called Sheila Brown, who owned a big house in Holland Park, and every time she walked into the kitchen he would go: 'Grrrrrrrrrrr. Eventually, she had to go in and assert herself: 'Okay Mr Pig, I'm boss.'

Mr Pig would often go and stay in Wallingford with my sister, Marian. He travelled there on his own and took a National Express coach. I would take him to Victoria coach station, pay his fare and say to the driver: 'I have a dog and he is travelling to Wall-ingford.' And Mr Pig would sit in the front seat looking out all the way contentedly. The driver did not mind and Mr Pig travelled like everybody else. He had this sensibility for looking very independent, and I can still see him on the bus with people looking at this dog travelling on his own. He would behave like it was none of their business. Somehow he would know to stay on until he got to Wallingford, and my sister would meet him in the market square. Marian would always call me as soon as he got there. Mr Pig was once sent home in disgrace from Wallingford because my sister decided to bath him. He was not crazy about the bath and he bit her hard. Absolutely furious, Marian phoned me up and just said: 'Be at Victoria coach station in two hours' time' and she put the phone down. So I went to Victoria coach station and Mr Pig slunk off the bus. I said: 'Mr Pig, you have blown it now.'

And sometimes I had to let Mr Pig go even when I wasn't away. Amelia begged me to let him stay for Christmas. Her grandmother was coming and she loved Mr Pig. I said to her: 'But you know, I'd like Mr Pig with me for Christmas.' But she said: 'Please, please let him come down,' and so I agreed. When her grandmother saw him, she just started crying because Mr Pig was there making his presence felt.

Mr Pig really excelled himself at my fashion shows. One time he had his own little stint on the runway with American model Kristin McMenamy and he also made his own impromptu solo appearances. He would have a little trek on the runway and would have a lie-down at the end, looking around at all the people out there. In fact, my favourite moment with

Mr Pig was after the fashion shows. At the end people came back stage, and with the TV crews getting their footage and your family and friends and supporters all saying 'hello', it can go on for quite a while.

Finally, when the last person had gone, the lights had gone down and it was quiet, suddenly in front of me was Mr Pig. He jumped up really high, offering his own congratulations in the sweetest way. He just kept jumping and jumping, and he never did that. It just got me right here, and I thought this sweet little thing knows what is going on and is offering me his congratulations in his own unique and most extraordinary way.

He met everybody, including Mick Jagger. We went to a party but I did not know him. There were only five other people in the room. He was there with his then wife Jerry Hall, along with Isabella and her husband. Suddenly I felt slightly awkward, but Mr Pig didn't feel awkward and was soon busy chewing Mick Jagger's foot. Mick said: 'What the fuck is this dog doing here? Get this dog off me.' I thought it really funny.

Mr Pig also had an interesting relationship with cats and was fascinated by them. Nick Rhodes had a cat called Yag and one day he swiped Mr Pig with his paw, and after that incident Mr Pig had a huge amount of respect for cats.

There were also some surreal experiences. Once, Isabella and Detmar (her husband) had gone away for the weekend and we were looking after their black pug called Alfie. Stefan and I were at home in Elizabeth Street and David Rocksavage, who is the Marquis of Cholmondeley, suddenly called up and said: 'I would love you to come and stay.' I said: 'I would love to but I have got two dogs.' He phoned back and said: 'Don't worry about the dogs, just go to Battersea heliport.' So when I got there, it was me, Stefan, Mr Pig, Alfie, Kate Moss, her daughter and David in this helicopter going to his house at Houghton. I thought Mr Pig would freak the minute the helicopter went up, but he behaved very casually and just sat there.

Mr Pig was so spoilt that he became accustomed to being treated as an extension of the family. I would go home of an evening and Isabella would be there. She would say: 'Good evening, Mr Pig'.

People would always say to him: 'Hello, Mr Pig' or 'Good morning, Mr Pig' and 'How are you today, Mr Pig?' He was very perceptive and clever and rose to every occasion. He had a sort of sixth sense about things.

Isabella had a show at the Design Museum, and dogs were not allowed in by the security men. So Alice Rawsthorne, the director, arranged for him his own pass, with his name and photo. When we pitched up with Mr Pig with his own pass, there was nothing they could say.

When I was given an award by Moët and Chandon at the V&A, I had to get special dispensation from the director of the V&A for Mr Pig to be allowed into the building. They were delighted to accommodate him because he was becoming more and more famous.

So much so that he had his own fan club in Japan, and Japanese girls would often write to him. I remember one day a Japanese girl came into the shop. When Mr Pig approached her, she started crying because she was meeting Mr Pig.

Wherever I would go in the world, people would ask about him. Once I went for a fitting with a member of the Spanish royal family and the first thing she said was: 'How is Mr Pig?' And I said: 'He is very well, thank you', but I was thinking to myself: 'How do you know about Mr Pig?'

I never ever got cross with Mr Pig because, even when he bit people, I felt they hadn't listened to his non-interest in them and that they had pushed it a little too far.

Anyone he didn't like I did not like either, and I always went along with his judgement. He was always right.

I remember we went to Heathrow airport to

collect Grace Jones, as it was her birthday. Mr Pig knew her very well but she had a new boyfriend with her. When Mr Pig saw her, she went up to him and said 'hello' as usual but the boyfriend thought he could do the same and ended up in A&E. Mr Pig bit him and there was blood everywhere at Heathrow. Grace said to me: 'I cannot believe that he did that. Now we have to go to A&E.' Of course, it was not great that this happened but Mr Pig bit all the right people. He was right and the boyfriend turned out to be not a very nice guy. Mr Pig had that sixth sense about people.

He also got the same feeling about a vet based in Elizabeth Street. The vet just treated him as one normally treats a dog, by picking him up and putting him down on the table. Mr Pig had never been treated like that. He was not used to being manhandled, and he just flipped out. From that day on, he would walk up the street and, when we would get to the vet, Mr Pig would look in the door and go: 'Grrrrrrrrrrrr.'

Typically, anyone that he did not like I was not crazy about them myself. I always trusted his judgement. That does not mean that I do not like Elton John, but I always listened to his intuition.

I remember a few years after he died, Grace Jones came round and she mentioned Mr Pig and said: 'I just have to say one thing. Mr Pig was difficult. I do not want to hurt your feelings but he was difficult.' Mr Pig was an exceptional character but I liked his difficulty because it made him more of a character, more of a personality. He was so gorgeous but he was not crazy about being stared at. When I travelled with him on the tube, people would look at him and he would look the other way.

On Christmas Eve, I always light a candle for him and put it on his grave because I miss him and think about him. When he died, I got nearly 100 condolence letters from people, including one from the Duchess of Cornwall. She loves dogs and knew what I was feeling. People were writing with genuine heartfelt sympathy.

I am still getting over his death because I will never forget him. When an animal is with you 24 hours day, you get used to them. Healing takes a long time.

DOG STORY

Juliet was Sir Edward du Cann's companion for 13 years.

CHAPTER 7

JULIET
Juliet and Me

EDWARD DU CANN

All the time that our children were growing up we kept dogs in our house in Somerset, in my Parliamentary constituency. You might say they were weekend dogs – I was busy in London all week attending Parliament, where I had substantial responsibility besides earning my living outside it – so I would be thunderously greeted by them late on Friday night when my constituency duties were ended for the day and we said our *au revoirs* again first thing on Monday morning.

Juliet came to me much later, after I had retired, and then we were together every day of the week. We had a very special connection that I never found with the five that came before her.

In Somerset first there was Watson, the Boxer dog. He was a close pal. I took him canvassing with me one day during a General Election campaign. Many householders greeted him warmly; one or two were terrified. Then the pub where I had lunch took objection to him even though he was wearing a large blue rosette. So that was the end of his political career. I couldn't be rude to the publican: he had promised me his vote. There are occasions, as any politician will tell you, when needs must.

Then there was Henry, a blue Great Dane. It was the breeder who said he was blue – he looked grey to me. My small son James used to wrestle with him on the carpet in the study. The dog was the invariable winner of these contests. My boy then became the owner of a King Charles Spaniel. In wrestling, he was now more often a winner than a loser.

When strangers used to drive by and stop outside our house, Henry the Dane would put his head through the driver's window of the car, to the terror of whoever had been foolish enough to have wound down his window. The rumour that we kept a monster was, we felt, a good deterrent to potential burglars.

And there was Cooper, the Retriever. If ever I took my gun out of the cupboard, he would go nearly berserk with excitement. When we were tramping across the fields, he always showed more energy than sense; but his enthusiasm was attractive and infectious.

Why did we give the dogs their names? Chiefly

because the family couldn't agree on what to call them. Some of their suggestions were daft, some hilarious and some were quite unsuitable. So we gave them the names of their breeders.

They were all wonderful, but the most remarkable dog we ever had was the latest member of the family. When we got her, she already had the unlikely name of Juliet. Juliet was a lady dog (I cannot bring myself to call her a bitch), longhaired, black and white, nicely marked. She was a Border Collie, born on a farm in Dorsetshire. To give her that name, the farmer must have been a romantic.

Before we married, my wife's family home was next door to the farm so she knew the farmer and his canine family well. And she knew Juliet's parents and saw Juliet many times after her birth.

Originally, my wife and her family had both Juliet and a brother from the same litter. The brother's name was Romeo. Had he stayed with us, we would have changed it. The pair of them would continuously encourage each other to hideous excesses, especially in chasing motor cars on the main road. To avert certain tragedy, the brother had to be found a new home and Juliet was alone with us. She didn't seem to mind too much. Perhaps she was growing up.

I have often wondered if she missed Romeo, her partner in mischief. Maybe she did for a little while but now she was settled in a new home with people who saw her all the time and who loved her. She seemed very content.

Juliet lived for a spell in Westminster until my retirement. Her exercise area was College Green outside the Houses of Parliament, where many interviews with Ministers and MPs are filmed by the TV crews. You may have seen her twenty years ago on the telly rushing about in the background of some shots.

On the whole she disapproved of cameras occupying the grassy area, which she liked to run across at enormous speed. Sometimes she would wander up to the interviewers and, barking away, give them her opinion. Some broadcasts had to be redone because of the noise of the heckling, but I was pleased that the television people seemed always to be good humoured about it.

After a bit of this, she would do her business in the bushes alongside the Embankment, confirming, as it were, her earlier barked opinion. She was fastidious, always neat and tidy and clean. Unlike other dogs who exercised on College Green, she never spent a penny on the Henry Moore sculpture, which the vulgar had christened Golden Molar. She respected Mr Moore's talent, if not everybody else did.

Then we lived for a while on the island of

JULIET

Alderney, an earthly paradise if ever there was one, assuming, that is, you like horizontal rain, fierce winds and sea spray on your windows after a gale (and there are quite a few of those). Juliet thought Alderney a paradise. She had masses of moorland on which to run and rabbits to chase – until myxomatosis came. Then, if she found one, she would stand like a Pointer showing me where there was a diseased animal. She never moved until I got the gun and shot it and put it out of its misery.

When my wife died of an incurable cancer, Juliet became my special friend. It was quite clear that she thought she had a duty to look after me. And she did: from now on, she never left my side.

We only fell out once in the years that she was my mistress. It was the day the Queen came to Alderney. I was one of a dozen members of the Royal British Legion who fell in at the War Memorial to greet Her Majesty and the Duke of Edinburgh. Our Chairman, a war time Royal Engineer Officer, swore Juliet in as a temporary member for the day. We legionnaires stood in a line. Juliet sat down quietly in the middle of it, the only one of us without a medal.

I had been in London the day before and had been one of thousands of the Queen's guests at her garden party at Buckingham Palace. It was blowing a hooley that day, but by chance on the day of the Queen's visit to Alderney the weather was marvellously sunny and calm. Our prayers had been answered.

When the Queen came to the end of the line to meet the ex-servicemen, Juliet stood up, as was polite. When the Queen came to me, she bent down to pat Juliet, and Juliet growled. I knew this was a sign that she was pleased, if not honoured, to have been noticed, but the growl would have sounded odd to some. So I was a bit grumpy with her. Not that the Queen seemed to mind; she understood dogs, of course. Juliet and the Queen had an easy rapport.

The citizens of Alderney knew how to act when there was a royal visit and how to make the most of it. When the little ceremony at the war memorial was over, the British Legion contingent boarded Alderney's bus and we motored down to the harbour, thereby augmenting the crowd that had come to see Her Majesty leave in a helicopter. That was our way to show our welcome: not to miss a minute of the visit.

The Duke of Edinburgh walked by, noting Juliet's presence. He said to me in the hearing of everybody: 'Edward, I see you have got yourself a guide dog at last.' Juliet was very pleased to be noticed again. She liked the compliment and she wagged her tail furiously, which I reflected was more tactful than growling.

Above: Mr Pig, a Jack Russell Terrier, accompanied Philip Treacy everywhere; from the cinema, to nightclubs, fashion shows and even photoshoots.

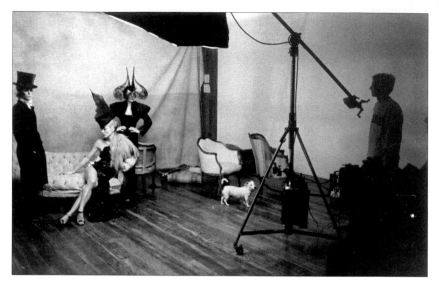

Above: Mr Pig always got involved in Philip Treacy's photoshoots. From left to right are Amanda Harlech, Daphne Guinness and Isabella Blow. In his time, Mr Pig was photographed by Mario Testino and Bruce Weber, and appeared on catwalks with top models such as Naomi Campbell and Honor Fraser. He had no idea he was a dog.

Below: Mr Pig and Philip Treacy being photographed with their great friend Grace Jones, who would sing to Mr Pig on his birthday. She also sang at his funeral.

Above: Juliet was a Border Collie born on a farm in Dorsetshire, and her brother was called Romeo; to have given them such names, the farmer must have been a romantic. Romeo went to a new home and Juliet became Sir Edward du Cann's devoted companion for 13 years, living out her final days in Cyprus.

Above: Lady Annabel Goldsmith promised her daughter Jemima a puppy if she passed her Common Entrance examinations. When Lady Annabel placed Copper in Jemima's arms, she fell in love with him on the spot. He was so small that she would sing him to sleep in her hands.

Above: Copper, a mongrel discovered in a butcher's shop, was Lady Annabel Goldsmith's constant companion day and night for 15 years, from 1983 to 1998.

Below: Copper was a talented and enthusiastic swimmer. He was in and out of swimming pools every summer, especially when he was younger.

Left: Harry, aged approximately 16 weeks, sips tea. This became a lifelong obsession. Tom Rubython found it was inadvisable to leave a cup within Harry's reach.

Above right: Harry with her mum and dad, Buttons and Dino. The photograph was taken on a return visit to where she was born to find her a little sister, Daisy, in July 2003.

Below right: Harry on her fifth birthday. It was the only time of the year she was allowed to sit at the table.

Left: Harry, two years old, sleeps in her usual place; a filing box on the editor's desk, where she ruled the roost until the arrival of Daisy.

Above: Harry (second left) enjoys a day at the seaside with friends Dee Dee and Chance and sisters Daisy and Doris. Wells, Norfolk, August 2009.

Above: Harry was a very keen, although not the fastest, swimmer in the engine pond at Castle Ashby, Northamptonshire.

Left: Harry setting out on her Sunday morning cycle ride through Althorp, together with half-sister Lila.

Right: The girls pose for their annual Christmas card, December 2008. The cards became a tradition and were much anticipated and prized by their recipients.

Above: A painting of Sir Edward du Cann's Border Collie Juliet. There was nothing sentimental about Juliet despite appearances to the contrary.

There was nothing overly sentimental about Juliet. She would only tolerate a single pat before walking off. She was an independent creature. If she came into a room into which I had entered before her and in which I was sitting down, she would check that I was occupied with a book or a newspaper or whatever. Then she would settle down a little way away, content to be still. But every now and then an eye would open to make sure I was not up to any mischief from which she might be excluded.

She was rarely demonstrative. If I went out to dinner and left her behind, when I returned she would get up, stretch, and seemed to be saying: 'Oh, so you are back again at last?' Nothing more than that.

If I was away for a day or two, there would be a bark and a wag when I returned, and that was it. Never any fuss. Juliet and I just took each other for granted; neither of us needed to be effusive. So long as I gave her her dinner, opened the door for her when she wanted out and was happy to have her with me on daily walks – to act as any gentleman would with his lady – she was content.

Then we moved to Cyprus. She had to have a microchip fitted by the local vet in Alderney, a kindly and good man. The visit to his surgery terrified her. I felt such a swine not taking her away before he could do his necessary work. It was out of character for her to show any fear. On the rare occasions when there were other dogs, even when they were vastly larger than her, she would stand no nonsense from any of them.

We cut her ample coat short because of the heat in Cyprus. She suddenly seemed much smaller. She was obviously happy in Cyprus. There were so many new smells.

Some of the native Cypriots, though normally warm and kind in their human relations, are not always as kind to animals as one might hope. Dogs that are used for hunting are sadly on occasion left abandoned in the countryside if their owners are dissatisfied with their performance in the field. Unless they are rescued and taken to one of the animal shelters that are manned by volunteers, Cypriots and British expatriates alike, they will starve to death. We have rescued our share of these animals. Also, many cats are semi-feral, and unwanted litters often require rescuing.

There was a goatherd who early in the morning every day took his animals past our house to their grazing place. Juliet would escape from the garden with the aim of keeping them in a proper herd. At first, the goatherd was distraught. He did not know what to expect from this strange creature, which looked quite unlike any typical Cypriot dog in the countryside. Then he got used to Juliet, and he and she and the goats became good friends.

It was a matter of astonishment in the small village where Maureen and I lived that, as we walked round with Juliet, we never put a lead on her; and even more, that if we called her to come she came and if we told her to sit she sat and never budged until we gave her a signal.

Herding was her instinct. When we were out walking, if Maureen and I parted, if one stopped to look at a vine or whatever caught one's eye, Juliet would not be content until she had rounded us up together again. Then off she'd go once more on one of her own little expeditions.

One of these proved perilous, potentially fatal. There are snakes in Cyprus. Not all are poisonous, but a few are. One day Maureen was walking with Juliet on a country track through the vines. They suddenly saw two snakes writhing together in some mating dance. At first, the snakes were oblivious of the walkers then one of them reared menacingly. Juliet moved at once to attack it. It was hard for Maureen to grab her and take her out of harm's way. Juliet never lacked courage.

With a name like Juliet one might have expected her to have romantic entanglements with other dogs. There was only one that I recall. A very good-looking stray roamed around the village for about a week. Juliet literally ran into him on one of her walks. She was coquettish, jumping about near him at one moment and running off the next. Then a stranger came and claimed him, and that was that.

It was a happy time in Cyprus but it was too good to last. Animals have such a short life span. She became frail. The hills were sometimes too much for her to climb. We would start a walk and she would lie down. There was nothing for it but to take her home and abandon the exercise. It was a kindness to take her to the vet to be put down. The day after we made the decision, she climbed on the back seat of the car in the way she so often did before our excursions and made herself comfortable. Maureen and I were so miserable. If Juliet sensed this, and she was a sensitive animal, she gave no sign.

When we came to the surgery, to my surprise, she showed nothing of the concern that she had shown previously in Alderney. The vet was a Russian lady and, like so many vets, she had the most kindly nature. Juliet smelled all the smells in her garden then let herself be picked up and put on the operating table. She lay there quietly, and then it was over.

We took her home. I had picked her up and laid her again on the back seat of the car. We buried Juliet in our garden and planted a cypress tree to mark the spot. I am happy to say that the tree is thriving. Now it is twenty feet tall and beautiful. You can imagine the thoughts that go through my mind every time I walk past it.

Just the other day, someone asked me how old Juliet was when she died. I am ashamed to say I could not say exactly. Dogs don't have birth certificates or death certificates even if they are micro-chipped. Since all my personal records, which I kept in the House of Commons, had been blown up by the Irish Nationalists, I never bothered to keep a diary. If I made a guess, I would say she was 13 when she died. I have always thought that an unlucky number.

I don't think I shall ever have another dog. You see, there never will be another Juliet and, if you ask them, everyone who knew her will tell you so – villagers, TV pundits, the lot.

In June this year, I saw a news item in an English newspaper that said that a pair of Collie dogs had been sold for £6,000 each. That's a lot of money. Juliet, as the family will tell you, was invaluable.

The heavy snowfall in early February 2009 delighted Harry but not her sisters, who did not emerge all day.

CHAPTER 8

HARRY
Harry's gone to heaven

TOM RUBYTHON

Just a few days before it happened, I was out walking my three Cocker Spaniel sisters Harry, Daisy and Doris on a beautiful summer's day. I started, for no particular reason, wondering what heaven would be like. Assuming it would be perfect, I asked myself what perfection would be – as I imagined it.

Any owner who loves their dogs is occasionally aware that they are likely to outlive them. As I watched my three, I imagined heaven 35 years hence, when I was likely to be dead but by which time Harry, Daisy and Doris would have been dead for at least 25 years. I imagined them all rushing towards me on a summer's day, at full speed across a field of long green grass.

Then it hit me: 'That is exactly what I have now'. I had already effectively gone to heaven, without going, and was experiencing heaven on earth – the real thing – every day this summer.

The old adage 'be careful what you wish for' could

not have been truer. Three days later, the eldest of the girls was dead. She was gone suddenly, and that dream of 'heaven on earth' was gone with her.

Owning three sisters, Harriet (Harry), Daisy and Doris, a year apart in age had come about almost completely by accident. It was not something I would have chosen, and in my heart of hearts I always worried whether I could get all three of the girls through to old age. The dangers they face to their existence from everyday life are the owner's responsibility when they take on the care of these creatures. But life became routine; and every day it lasted, the more routine it became. And the more routine it became, the easier it was to take for granted.

When the end came, it was sudden and unexpected and a tragic accident. It all started on a Monday morning. The morning of 28th June 2010 was a morning like any other. After their first walk at 7am, we went to the old coal wharf, latterly the boatyard, where my office is and I reversed into my usual parking spot. I opened the car door and the dogs rushed for the exit. I counted out the three of them as normal. They were so impatient that they always jumped out of the car over my lap. I always counted them in and out ever since Harry had been mistakenly arrested by the Metropolitan Police in London and taken to Old Street Police Station. She had

stubbornly lingered by the entrance to an underground car park on the way home, sniffing out the
hope of a juicy morsel from the dustbins of the
Chinese restaurant next door and was taken in by
the police for loitering – or the dog equivalent. It
was a few hours before she was returned to me; the
officers, having been called to an emergency on
the way back, decided that it would be more expedi-
tious to take Harry along with them, sirens blazing
and blue lights flashing, and to bring her home later.

She returned having made new friends and looking
rather pleased with herself. It was just one of her many
adventures.

Harry was a participative dog in every way, so
getting in and out of cars was second nature and
routine; she was so used to travelling.

That morning, I opened the door and went to the
boot to get my bag. I left the car door open as I
always did. I shut the boot and then the car door.
The process took at least a minute as I shuffled in
the boot looking for everything I needed after the
weekend.

But one thing was different that morning: Harry
had sneaked back into the car and was now lying
in the back seat, seemingly figuring that the car,
still cool from the air conditioning, would be more
comfortable than her basket that morning. It was

totally out of character and something Harry never did. Harry was the sociable one; she was far too interested in meeting people and getting involved in the daily routines of the four or five businesses that occupied the site, known as Billing Wharf.

I didn't bother to check the car again because I could see Daisy and Doris walking towards the office door. It never occurred to me for one minute that Harry might still be in the car. It was her habit to rush off to the river bank, so I thought Harry must already be by the river hunting for a stray piece of bread the swans and ducks had missed. Or perhaps she was visiting her regular haunts before shuffling over to my office later on. As always, Harry had her own agenda and would be pursuing it.

Getting back in the car was something she had never done before in the 8,000 car journeys I calculated I had made with her over the past eight years. Sure, she would jump in a car in advance of leaving, but she never liked to be left behind. She was too interested in what lay in the day ahead and, in particular, in everyone's breakfast and lunch.

My Dad came into his office next door and he didn't notice Harry's absence either. Harry would often spend hours with him, anticipating the arrival of his lunch (which meant her lunch too).

I rose from my desk to take the dogs out for a walk

by the river and for a break in the writing, which that day was not coming as easily as in others. I called the girls and Daisy and Doris came to the door but not Harry. So I called out for Harry and she didn't come. In that instant, it suddenly dawned on me that I hadn't seen her for a while. And that is when I suddenly realised something might be wrong. Then, as quickly as it takes, I wondered aloud whether she could be still in the car. Even then I didn't think it possible. The car was equipped with motion sensors that should have set off the alarm at the slightest movement. Nonetheless, I ran over to the car as fast as I could and there she was slumped on the rear seat, looking down at the floor. I banged on the glass but she didn't move. Feeling my pockets for the keys, I realised they were on my desk and I ran back to the office and shouted to Carly and Sophia, the girls in the office, that Harry was still in the car. By then I feared the worst, and they must have as well. I ran back over to the car and unlocked it. I threw back the front seat and touched Harry. But she was motionless.

I screamed out loud: 'No, not Harry. Not my Harry.' I always called her 'my Harry'.

But at that moment I knew Harry was gone, and gone forever. It didn't matter to me that she was dead – what mattered was that I would never see her again. That feeling just swept over me – that things would never be the same again without Harry around.

I lifted Harry out of the car and carried her to a bench by the river and just sat there for the next hour and a half with Harry in my arms, crying, repeating her name in between apologising to her for not keeping her safe.

I had always been pragmatic and prepared myself for their deaths, and especially that of Harry as she was the eldest and the least fit. But what I instantly realised was that I had not prepared myself for (and not even thought about) the fact that I would never see her alive again – that thought overwhelmed me.

In that moment, I became overcome by the feeling that all the things I took for granted with Harry would never ever happen again and that somehow I had let her down and not kept her safe. It was a truly desperate feeling. I would have given anything for Harry to come back to me.

Harry was a special dog, and special dogs do special things every day. It was unbearable to think she would never be doing those 'special things' ever again. Everything that had happened over the eight years of her life had served to create and consolidate the love and the bond between us. It was suddenly gone and, for now, I desperately wanted to be close to Harry while she was still warm and there was a sense of her being alive.

I knew the next 24 hours before we buried her would be the time to remember her and to say 'goodbye'. And for those 24 hours, although she was dead, I felt that her spirit stayed with me.

In fact, the only sign that anything was amiss was her little pink tongue clamped firmly between her teeth, protruding out of her mouth very slightly and dry at the end. Otherwise, she looked normal and very, very serene.

I just held Harry tight in my arms and talked to her the entire time. Every feeling and every emotion I had flowed into that dog. Daisy and Doris just hung around by the bench, obviously sensing something awful had happened but not knowing what. Eventually my Dad came over; he was almost as distraught as I. We wanted to be sure Harry was dead as she was still so warm and looked so normal. But there was no sign of any movement.

Later, I drove Harry home for the final time. It was just the four of us – Harry, Doris, Daisy and I – with Harry lying on the front seat of the car just as she might have done had she been alive.

At home, I placed her in her basket and put it on the kitchen floor. I lay her in it with her head on the side. She looked so alive, and her warmth perpetuated the illusion. As she lay in her basket, it seemed still the same old Harry. As I made the girls' tea, I made three plates as usual and put the third one down by Harry's side.

I called Ania, Harry's 'mother', and arranged her burial for the next day at three o'clock. It was then decided that Harry would be laid to rest in the middle of Ania's lawn, underneath a quince tree, where she could rest in the shade.

To ease the grief, I took the two girls on a long walk through all of Harry's old haunts. We walked and walked until the late evening heat subsided. The girls, searching for water, found themselves immersed in a stagnant pond they hadn't discovered before and were caked in black, smelly mud which necessitated 30 minutes in the bath when we returned home.

Still warm, Harry's eyes were so crystal clear I wondered whether she might suddenly jump up.

I was determined to have one last night together with her, as normal as possible. So I lifted Harry upstairs from the kitchen and put the basket on the end of the bed exactly where Harry would have spent the night had she been alive. By now, a little bit of blood was dripping from her nose. Apart from her tongue, it was still the only sign she was dead. It was as if her body was only slowly beginning to shut down. In fact, she remained warm until the next morning and her eyes were still clear.

The following morning, I talked to her and sensed she was still taking it in. I carried her down to the kitchen and made the breakfasts as usual. It seemed wrong and strange not to make three portions. I put Harry's down beside her basket. By now, Daisy was off her food completely and clearly unsettled by what had happened. After breakfast, I carried Harry into the sitting room and put her on her favourite settee and went out with Daisy and Doris for another long walk, remembering Harry as we walked. Before that, I arranged with Mick, a friend and long-standing working colleague, to dig Harry's grave.

When we returned, at around midday, Harry was cold. Her eyes were misty and her soul had finally departed at around 11 o'clock, I guessed. Harry's sense of timing was always good; I could never have buried her if I thought she was somehow still present and hanging around.

By the time we finished, it was 2:45pm. I placed Harry's basket in every room of the house where she had spent time, ending in the kitchen. I then took Harry out to the car for the last time. Daisy and Doris were already on board. Harry was on the front seat and her sisters in the back.

The last car journey was tremendously emotional, and I wasn't really ready for it. Every day for eight years, Harry had jumped in and out of my car. I had thought nothing of it, and now it was to be the last time.

I realised that there will always be a 'last time', whenever it occurs. We passed by the office and I took Harry back to Park House, where she had lived previously and had been so happy, and then to her final resting place at Autumn Cottage, Ania's house, where she had lived for two years before I moved to Castle Ashby. It seemed right that she return to the village where she had spent most of her life. There was a sense that this was what she would have wanted. The journey took three quarters of an hour. Upon our arrival, Ania, my Dad, my stepmother Margaret and Mick were all awaiting Harry.

Mick carried her up the stairs to Ania's garden, where he had dug a deep, wide hole big enough for her basket. We placed in it some of her favourite toys and some photographs of her and her sisters together. I removed her collar, and gently we lowered her into the ground. When she was in, I kneeled down and kissed her for the last time. Then we covered Harry with a soft fleece to protect her and covered her with earth, and that was it. Harry was gone.

Harry was born on the 7th July 2002 in a small village in Worcestershire. The house was a red brick lodge with crenellations along the top that gave it the appearance of a castle. It was appropriate, as

HARRY

Harry was a queen amongst dogs.

Ania had found Pat, her breeder, on the internet. It turns out the dogs (and cats) she breeds are of a perfect temperament. It made me realise that breeding isn't merely down to chance – it is truly a skill.

How these dogs are treated when they are born, and the atmosphere in which they spend their first weeks, is so important in how they turn out. When I looked around Pat's garden, I saw happiness and perfect animals. I also looked at her children a boy and a girl. They were perfect too. It was a surprise, as the house and garden in which they lived were absolute chaos.

Over the years, another 14 dogs and a few cats from Pat went to various friends and relatives and, without exception, they each brought great happiness to their respective owners.

Harry arrived with me in the third week of September as a birthday present from my then new girlfriend, Ania. Had I been given the choice, I would always have objected to having a dog, especially living in a third floor flat in London with no lift or garden.

There are so many reasons to say 'no' and there is never a right time. When Ania asked me whether I would like a dog, I said: 'Don't do anything stupid. How do I know if I will like the dog?' But somehow, and against my wishes, Harry arrived.

Ania takes up the story: 'She was one of six sisters in the litter and only two were available. They all looked identical in the puppy pen. Had they been male dogs rather than bitches, I was told, they would have all been sold.

'I chose Harry because at the time she was the whitest. I was wearing trainers and, when I put her on the ground, she pulled at my laces as if she wanted me to choose her. She looked like a little white piglet with the markings of a Friesian cow. I thought the patch on her back was heart-shaped, which I rather liked.

'I agreed to collect her after the Italian Grand Prix, as that was the last time that year that I would be away. When I brought her home, it was in my Mum's cat basket. We stopped in Stratford-on-Avon where my Mum lived. At that time, I had no idea of what to do with a dog; how to train it or otherwise. Harry frightened me as she was so tiny, and even when she got the hiccups it was terrifying; I thought she was dying. I remember she clamped her teeth around a magazine rack at my Mum's house that first day and, apart my fingers, that was the only thing that she ever chewed, even though people warned me that Spaniels were partial to chewing up things.

'She came with me to Harlestone later that day. I had an appointment with an estate agent to view

Park House, which would be our home for the next five years. Afterwards, I took the train back to London to give her to Tom. Harry insisted on sitting on my lap in the train and was very good the entire way, until she peed on me as we pulled into the station. It was very embarrassing when I handed her over to Tom all wet.'

So when Harry arrived on the train from Northampton to Euston there had been no plan. But how could I not like Harry? Harry was mostly a white colour then, with flashes of black. She was so small she could easily go in my pocket, and often did.

She was absolutely no problem in the car, and when we got home I carried Harry up and she began to explore. Of course, she wasn't called 'Harry' then. I had no idea what to call her. 'What do you call a dog?' I simply had no idea; it was all so new to me.

That evening I took her out onto a patch of grass on the Old Street roundabout in London, as I had no idea where the local parks were. She immediately disappeared in the grass, which wasn't particularly long, and I wondered if I had lost her already. After five minutes, I spotted her in the dark and took her home in my hand. A friend stopped me by the door and, even after a ten-minute conversation, he failed to spot Harry in my pocket – she was that small.

Harry was immediately at home in Old Street and spent the night downstairs in her basket. Someone had told me to establish that discipline and I remember thinking: 'This is too easy'. In the morning, Harry was straight in the car as we drove to the office and she spent much of the day in a basket by my desk, coping with the inevitable attention from the 50 or so people in the office by falling asleep.

Still there was no name, and Ania started calling Harry 'Thomasina'. I wasn't fond of the name so we had a vote amongst the staff in the office, but nothing seemed to fit. I racked my brains until eventually 'Harry the Hound' came into my head and I thought: 'We'll call her that.' Later her name was changed to 'Harriet' for formal purposes, as she was a girl, but she was always known as Harry the Hound after that.

By the third night at home, Harry had worked out that I didn't sleep downstairs. Wanting to know exactly where I slept, she made it quite clear that downstairs was out and that henceforth she was sleeping with me. Eventually, she settled herself on the end of my bed and began a routine that would endure for eight years. Harry slept where Harry wanted to sleep, usually on my pillow and, for a small Spaniel, quickly became an immovable, heavily-snoring object.

As Harry grew up, the change of status prompted me to rent a house in Northamptonshire, in the village of Harlestone, for weekends. So we began a routine of weekdays in London and weekends in the country. At first, she travelled by train to and from London. She was so friendly and endearing, she would often find a welcoming lap amongst the commuters. Eventually, Harry quite happily got in the car to go to Park House every Friday, and then the reverse at 4am on Monday mornings. There was never any question of her not coming.

She wanted to come everywhere. She could never be left behind; it was always: 'Me too'. She loved the car but always wanted to be in the driver's seat. It was virtually impossible to convince her otherwise. Eventually, she would settle for the warm lap of a passenger but, once settled there, Harry would not be moved.

Early on, I found a park in London called the Bunhill Fields, situated amongst ancient gravestones, and Harry travelled there in my pocket. She pottered quite happily in the grass twice a day whilst I read the paper. After half an hour, I popped her back into my pocket for the journey home.

Ania took her to puppy-training classes in London, as she recalls: 'I was told to come early. They warned me that she would be frightened, but assured me not to worry and to leave her. The other dogs were huge compared to Harry: a Labrador, a Border Collie and a variety of mongrels. Unexpectedly, my tiny puppy chased the pack of dogs around the room, barking fearlessly. The trainer, who previously had said that he did not like small dogs and only kept Rottweilers, offered to buy Harry on the spot. He told me she would not make the best family pet, as she was so headstrong. He was both right and wrong; she was headstrong but a wonderful pet and not hard at all.'

As the first year rolled by, Harry became more and more loved. In October, I left my job editing the magazines and it coincided perfectly with a move out of London and to the country.

Park House was the ideal place for dogs, but for its proximity to a large adjacent field of sheep. Like all Cocker Spaniels, Harry loved to chase sheep. Fortunately Ivor, the local farmer was very understanding with this tiny naughty puppy. But I was determined to put an end to the sheep chasing and, after two months, Harry responded to a firm hand and learnt to walk through and around the sheep without paying them any mind.

It was made clear to Harry, by spells in solitary confinement in her basket in the kitchen, that she had done something wrong. Dogs love you all the more when they know the order of things. But before that, of course, Harry took advantage. As Ania recalls: 'Occasionally I would see a little flash of black and white going down the drive and out of the front gate into the field, scattering the sheep. I would have to run out in my pyjamas and dressing gown to get her back. Having just moved in, I wondered what the neighbours would be thinking.' When she finally got the message about the sheep, Harry's determination meant that, rather than take the longer route around the field, she would walk as slowly as possible through a flock with her nose to the ground, oblivious while the sheep parted like the seas for Moses. This became a pattern with Harry; she only ever walked slowly and far behind – nothing could hurry her.

She loved people like no other dog you can imagine. If you liked Harry, she loved you and so began a series of stopovers where Harry would visit other families and stay the night.

When I eventually returned to London in the spring of 2003, Harry came too. We returned together to the country at weekends.

At the start, Harry was an only dog and there was no intention for her to be anything else. Cocker Spaniels tend to form a bond with their owner, one person, and that is what they are accustomed to. Once established, that bond is hard to break.

Harry also developed several very individual habits that defined her character. She was so determined to be her own dog. She was very particular about the door through which she entered the house. No matter how many other doors were open, she would enter only via the kitchen door. If the door was not completely open, she would stand at it and bark. She would never ever push a door open with her nose, as she was far too grand a lady for that.

Althorp is only a mile from Park House. When Princess Diana died and Earl Spencer opened up the house and museum devoted to her, the back road to Althorp House was closed. Over two miles long and surrounded by countryside, the road became the perfect dog walk. But it was at least a mile away, so I got a bicycle with a pannier bag in which Harry was given a lift to her walk. She became the talk of the village with her head poking out of the pannier bag.

As an only dog, Harry revelled in the attention. She understood she was the focal point of the household and took full advantage of it. For instance, after a cup of tea wherever we were, Harry would move in for the dregs. She always drank tea out of the cup and showed her displeasure when all the tea was drunk up and nothing left in the cup for her.

And she never waited to be offered. She had her

favourite snacks and would have been a happy vegetarian dog, who also liked meat. She always wanted raw vegetables as they were being prepared; she was very partial to sprouts, which would be cunningly removed from the fridge should the door be left open and a back turned. A new acquaintance would be immediately tested to find out how soft a touch they were. Having given in once, their card would be marked by Harry and she would then stamp her feet from side to side until she got what she wanted.

When not offered, she helped herself. She stole the baby pumpkins off the plants in the garden so that there were hardly any to pick. When Ania tried to plant asparagus, Harry ate the crowns out of the bucket in which she had been storing them. On walks, when she smelled a barbecue, she would sneak off to find it. She would inevitably be delivered back by car after having joined a family cook-up in the village.

But her greediness didn't always work out. Harry loved eating the red rubber bands that postmen in London tend to leave everywhere. Her sneakiness knew no bounds and it was virtually impossible to stop her. It never seemed to affect her too seriously though. The bad habit ended only when I left London in 2009.

Once, she ate an entire box of chocolate truffle hearts wrapped in red foil. She ate the foil as well and then pooed it out just the same as they had gone in – wrapped in the red foil.

Harry also had an amazing memory, as Ania says: 'She was once given a Jumbone by my Mum and she hid it under mum's pillow. Mum left it there, and on a return visit many weeks later, Harry went through the front door and straight up to the bedroom to check it was still there – which it was.'

Harry became a very sociable dog. She made friends with everybody but, despite her dalliances, she remained devoted to me.

And that is how it would have stayed but for the summer of 2003, when my Dad said to Ania to choose him a relation of Harry's for his next dog. His own dogs were getting old and he did not think they would survive too much longer. Ania didn't need much encouragement to make a return journey to the breeder and, exactly a year later, Daisy, Harry's half sister, arrived.

Daisy was going to stay with us initially and go to my Dad when his dog died. But it didn't happen like that. My Dad's dog suddenly revived and got a new lease of life and Daisy stayed with us. Eventually, she too wanted to come to London in the car every week.

After that, Harry was no longer an only dog. And Harry subsequently changed. Daisy was a different character, far more needy and less confident, but just as determined. This time we were resolved that she would learn to sleep in the kitchen in her basket. But after two nights, she realised Harry was sleeping elsewhere. And that was it – Daisy caused the most almighty stink until she was given equal status with Harry.

She also decided that, as I was to be her master, she would have to initiate a campaign to ease out Harry. There was no way Harry was going to challenge Daisy so there began a new chapter in which Harry became everybody's dog. Whilst Harry came with me to London, she was still top dog, but eventually Daisy made her way there too by demanding to come with us. Harry stoically accepted the new order. Eventually Daisy became so settled that it was obvious she wasn't going anywhere.

Then, a year later, precisely the same thing happened to us again, albeit with a twist. A visit to the breeder with friends established that there were three virtually identical sisters available. These three gorgeous puppies were only distinguishable by size: large, medium and small. Unable to separate them, we brought home all three. The idea this time was that my Dad, now dogless, would have two of them, and the third would go to my brother.

So Doris, Delilah and Dolly all arrived – three of the most gorgeous bundles of fluff you could imagine. Doris was the larger sister, with Delilah in the middle and Dolly the smallest. These were Daisy's sisters and Harry's half-sisters.

Harry and Daisy did not know what to make of this sudden expansion of the family. On walks, they trooped along, all off their leads, and all perfectly disciplined together. The three little ones just followed their two elders. I remember coming across a lady who said to me: 'Ooh, you have a fleet of Spaniels.'

After the sisters arrived, we installed a dog flap. As Ania recalls: 'Harry was the last to use the dog flap, and I never thought that she could use it until, one day, after having locked the house, I found her behind me. I thought I had locked the door so I checked and found her again, I did this twice. Eventually, I hid around the corner and spied her coming through. I never thought that it could happen.' One of the sisters was supposed to go to Chris and Ann Cleverly, friends in Plymouth, but when push came to shove I couldn't separate the three sisters over Christmas, and Chris and Ann eventually acquired a puppy from the same breeder. Chris took the dog to Plymouth where that dog, Holly, has lived happily ever since.

All five dogs came to London every week, and I remember all of them being in the pockets of my Barbour jacket with Harry and Daisy walking alongside to Bunhill Fields cemetery.

Doris and Dolly were duly delivered to my Dad's house for Christmas. In truth, he and my stepmother Margaret couldn't cope with two lively Spaniels, and after Boxing Day the two of them came home again. We immediately found a home for Dolly with a local solicitor and his wife. Delilah went to Philip and Wanda, an almost-retired couple who moved to the New Forest. They later went back to Pat and bought Wilma, Harry's sister and Doris' and Daisy's half-sister. Wilma is another incredible personality.

Doris stayed and joined our family full-time. Now there were three girls together, and Harry had to adapt again.

Two years ago, I split up with Ania and the four of us moved to Castle Ashby. I had also left London again in 2009 after selling my magazine to new owners.

Although Harry only lived in Castle Ashby for a year, she became well-known due to her insistence on joining all social occasions. The girls, although technically banned from the club, were welcome at

The Falcon, a local pub nearby.

Castle Ashby is a very sociable village with a Friday-evening-only club. But Harry was banned from the club by the committee (Castle Ashby is made up of dog families and cat families, and the cat families are always manoeuvring against the dog families). I was the only dog/cat household and consequently the only one (I thought) with any proper perspective.

Anyway, one night, after Andy, the chairman of the committee, ordered Harry out of the club, she retired to The Falcon, where she was always welcome. Harry joined and sat down with two bikers, who were outside enjoying pints. Every so often, I would poke my head out the door and see the two bikers sitting on the grass and Harry sitting opposite them engaged in some sort of communication. Eventually the bikers departed and Harry sensed it was safe to sneak back into the club. She was right; the chairman had gone and she could roam safely amongst the members. It was so typical of Harry.

A later committee meeting upturned the ban and dogs were welcome at the club on leads. As the sisters hardly ever had on their leads, this was not an option.

Daisy and Doris resigned themselves to the fact that their Friday-night socialising had been curtailed. But Harry would not be thwarted. On Friday nights, when I was leaving for the club, she would follow me by breaking through the barricaded dog flap. When I put her back in the house, she would follow me again. When I put her back a third or fourth time, she would wait half an hour and then join me later on site, making the five-minute walk on her own.

Not actually knowing where I was, she would check The Falcon first and then the club. Her sense of pleasure when she found me was profound. She loved the club but if I happened to be elsewhere in the village, she would always find the house and arrive, barking at the door for admittance.

Likewise, whenever I took the car, Harry would run up the verge after me until I stopped and let her in. She wouldn't care that her sisters had been left at home. And so on it went happily.

But on Harry's last weekend, she started behaving differently. That Sunday afternoon, I visited Ania's on my way to my Dad's house. Only this time, Harry didn't rush to the car when it came time to leave. I called her and she wouldn't come. She had chosen to stay with Ania and slept by her side on the sofa in front of the television. When I returned two hours later, she dutifully got in the car.

You don't notice these things at the time but, afterwards, I sensed she had stayed on to say 'goodbye' to Ania.

Then, on our walk later that evening in Castle Ashby, Harry hung back, as she often did, but five minutes later when I had reached home, she galloped down the road for quarter of a mile at full speed and arrived with the hugest of smiles on her face. Harry never ran and always trundled in at her own pace.

But I didn't pay much attention to either of the changes in her routine at the time.

After she died, recalling events, I got the strong impression that Harry had a sense on that Sunday that she was going to leave us. She had visibly aged in the last six months and now, looking at old photos, I hadn't realised how much. She had become very grey around her nose and her eyes were beginning to go cloudy. Her final run down the road and her farewell to Ania strengthened that feeling. I believe Harry got back in the car to die, and that it was her wish. But I'll never know.

Daisy and Doris have adjusted to Harry's absence and life does go on, but things will never be the same again without Harry, who started it all and brought so much joy into so many lives, and who has definitely gone to heaven.

DOG STORY

APPENDIX
A history of dogs

2100 BC
The first domestic dog – the Saluki

Saluki dogs were highly regarded in Egyptian culture and are considered to be the first-known domesticated canines, with their images appearing in tombs from 2100 BC. They were so highly respected that in some instances they were mummified alongside the bodies of the Pharaohs.

1104
The first (and only) dog is appointed to rule a country

King Eystein of Norway announced to his people that, as punishment for the death of his brother, either his dog Saurr or his slave would rule the country. The people chose Saurr, believing he would he would be gone sooner than the slave. During the three years of his rule, it was rumoured that Saurr signed decrees with his paw print.

1796
Tax on dogs is repealed

A tax which was levied upon working dogs with tails was repealed. The tax had contributed to the widespread docking of dogs' tails since Georgian times. Historically, tail docking was thought to prevent rabies, strengthen the back, increase the animal's speed and prevent injuries. The practice persisted, albeit on a much smaller scale.

1822
The Martin's Act is passed in parliament

The Martin's Act was the very first animal welfare law, and it forbade 'the cruel and improper treatment of cattle.' 13 years later, in 1835, it was renamed the 'Pease's Act', which consolidated the existing law and extended the prohibition of cruelty to dogs and other domestic animals. At that time, bear baiting and cockfighting were also made illegal.

1824
The RSPCA is founded in a
London coffee shop

The world's first animal welfare charity was originally called the Society for the Prevention of Cruelty to Animals (SPCA). One of the founders in the coffee shop was William Wilberforce, the slave trade abolitionist. The charity gained royal patronage in 1837 and Queen Victoria gave it permission to add the royal 'R' to its title in 1840, making it the RSPCA. Today, it is a worldwide organisation. Charles Dickens was an early supporter.

1859
First organised dog show is held in Britain

The first organised dog show was held in the Newcastle-on-'Tyne town hall on 28th and 29th June 1859. The show was organised by a Mr Shorthouse and a Mr Page at the suggestion of a Mr Brailsford. There were 60 dogs entered, of the Pointer and Setter breeds. One class was held for each breed, and dogs were unidentified except for their kennel names. In the final, a dog named Spot, owned by Mr Murrel, competed against Venus, owned by Mr Brown, for a prize of 22 shillings (equivalent of £1.10).

1870
Kennel Club is founded by a
Member of Parliament

A controlling body to legislate in canine matters was founded by Mr S Shirley, the Member of Parliament for Warwickshire. He called it the National Dog

Club Committee. The first committee meeting was held at 2 Albert Mansions, Victoria Street, London, on 4th April 1873, with 12 members recorded as present. This meeting effectively marked the founding of the Kennel Club.

1880
Dog registration system introduced

The Kennel Club introduced a system called 'Universal Registration', initially opposed by members, for reserving the use of a name for a dog. The system quickly gained wide acceptance and opposition melted away. Registration was intended to avoid duplication in the Stud Book. At that time, a pedigree was of little importance and only came as an aid to identification.

1898
Cropping of dogs ears is banned

His Royal Highness the Prince of Wales leads a campaign to ban the cropping of dogs ears and, from 9th April 1898, dogs with cropped ears were deemed ineligible for any form of competition under Kennel Club rules. The practice effectively ceased altogether.

1905

The first canine film star makes his debut

The first canine film star appeared in *Rescued by Rover*, a short silent movie directed by Cecil Hepworth. After its release, the dog, known as Blair, became a household name. It is said that the film was so popular that Hepworth had to record it twice due to the negative wearing from the high demand for prints.

1914

Dogs are recruited for military service in World War I

Various breeds of dogs were recruited for active service during World War I, including Collies, Sheepdogs, Lurchers, Welsh and Irish Terriers and Airedales. They played a vital role not only as messengers in the trenches but as a great comfort to the soldiers who were experiencing the horrors of trench warfare.

1924

Thomas Hardy writes an ode for the 100th anniversary of the RSPCA

On 22nd January 1924, Thomas Hardy wrote an ode called 'Compassion' in celebration of the centenary of the Royal Society for the Prevention of Cruelty to Animals. It contained three verses.

1925

Siberian Huskies deliver antitoxin on a 674-mile journey

Twenty mushers travelled 674 miles in order to deliver a vital antitoxin to the town of Nome after a deadly outbreak of diphtheria spread through the town's young inhabitants. A Siberian Husky named Balto led a team on the final leg of the Great Serum Nome run. The run was successful, and Balto gained great popularity and praise. His escapades were even made into an animated film produced by Steven Spielberg's animation studio.

1931

Tobey becomes the wealthiest dog in history

A Poodle named Tobey became the richest dog in history after being left £15 million by former owner Ella Wendel in her will. The money has subsequently been passed down through a line of privileged Poodles, with the current successor, Tobey Rimes, boasting an inheritance that has grown to £30 million.

1934

A statue is erected in honour of the loyal dog Hachikō

A bronze statue of Hachikō, an Akita, was erected at Shibuya station, where the dog waited loyally for his deceased master to return every day for nine years after his death. Hachikō's faithfulness became a symbol of loyalty in Japan and, each year on 8th April, his devotion is remembered in a ceremony at the station.

1936

Faithful Old Shep begins his five-year vigil

Old Shep, the 'forever faithful' dog, began his five year vigil at Ft Benton train station. His owner, a sheep herder who remains nameless to this day, fell ill and was taken to hospital. Shep waited outside but, unfortunately, his owner was not to return. His coffin was taken to the train station and loaded onto a train. Shep followed and waited patiently for his owner's return for five years, greeting each of the four daily trains at Ft Benton. He was featured in Ripley's 'Believe it or Not' and became well known in Depression-era America.

1940

Robot the dog discovers
17,000-year-old paintings

Robot the dog discovered Paleolithic cave paintings, estimated to be around 17,000 years old, in Lascaux, Southern France. Whilst on a walk with his teenage owner Marcel and his three friends, Robot got lost and fell into a cave. Whilst rescuing Robot, the boys unwittingly discovered one of the most important sites of prehistoric man.

1942

'Lassie' makes its big screen debut

The first film featuring the famous character of Lassie the Collie was released. It was titled *Lassie Come Home* and the lead was played by several male dogs due to the fact they were considered to be more attractive than female dogs. Also, with a child acting alongside them, the male dogs appeared larger and the child smaller, which was thought to be more desirable in the film.

1957

Laika – the first dog in space

The first dog, and arguably the first living being, was launched into space on the Russian Sputnik 2. Laika orbited the earth successfully but sadly perished as the craft burnt up on re-entry into the atmosphere. It was later revealed that the craft had not been designed to make a safe landing. Much controversy ensued, with animals rights protesters expressing outrage at the Russian government's misleading statements before and after the flight.

1965
John Noakes joins 'Blue Peter'

John Noakes joins Christopher Trace and Valerie Singleton as a presenter on *Blue Peter* on 30th December 1965. It is a golden era for the children's television programme. The original *Blue Peter* dog was Petra, who made her debut in 1962. Patch was the son of Petra and was looked after by Noakes when he joined the show.

1966
Pickles the dog discovers the stolen Jules Rimet World Cup

Mr David Corbett was walking Pickles when the dog's attention was caught by a bulky package that had been wrapped in newspaper and placed under a bush in someone's garden. Lo and behold, it was the Jules Rimet World Cup, which had been stolen from under the noses of football authorities whilst being exhibited in Westminster prior to the start of the tournament. Mr Corbett received a £5000 reward.

1971
Britain's most famous dog debuts on 'Blue Peter'

After the sudden death of Patch in 1971 from a rare disease, *Blue Peter* presenter John Noakes is given a Border Collie puppy, christened 'Shep' by viewers. The excitable puppy is told to "Get down, Shep" by Noakes and it instantly becomes Britain's most famous catchphrase overnight.

1975
Political leader brought down by a dog called Rinker

Liberal Party leader Jeremy Thorpe became the first British politician to be forced to resign by a dog. Norman Scott, a former male model, was walking in Exmoor in October 1975 with his Great Dane called Rinker. Andrew Newton, a former airline pilot, armed with a gun, shot and killed the dog and then pointed the gun at Scott, but it failed to go off. Thorpe was accused of being behind the attempted murder but was acquitted.

2001
Guide dogs lead the blind from New York's Twin Towers

When terrorists attacked the World Trade Centre in New York in September 2001, guide dogs Riva and Salty calmly led their owners down smoke-filled stairs from the 71st floor. They escaped the building safely just before it collapsed, and Riva and Salty were later honoured for their bravery.

2005
'Marley and Me: Life and Love with the World's Worst Dog' is published

The autobiographical book, written by journalist John Grogan, was a *New York Times* best seller and chronicles a thirteen-year period in which Marley, the family's boisterous and destructive Labrador Retriever, lived with them. The story touched many people's hearts and Marley's eulogy, written by Grogan, received more critical acclaim than any other work written during his professional career.

2006
Tail docking declared illegal in Britain

In March 2006, an amendment was made to the Animal Welfare Bill that makes the docking of dogs' tails illegal, with the exemption of working dogs, such as those used by the police force, the military, rescue services and pest control. The amendment was supported by a majority of 476 to 63. Docking, previously standard procedure for some breeds, was now punishable by a fine of up to £20,000 and up to 51 weeks of imprisonment, or both. In Northern Ireland, the practice of tail docking is still legal.

2008
'Marley and Me': The movie

The book *Marley and Me* is turned into a film starring Jennifer Aniston and Owen Wilson, and directed by David Frankel from a screenplay by Scott Frank and Don Roos. The film opened in the United States on 25th December 2008 and set a record for the largest Christmas Day box office ever with US$14.75 million in ticket sales. Overall, it grossed $143 million in the USA and almost US$100 million outside of America, for a total worldwide box office of nearly US$243 million. It is believed to have done as much again in DVD sales and turned a profit of over US$100 million for 20th Century Fox, the studio which release the film.

2009

America has dog population of 77.5 million dogs

On 30 December, the American Pet Products Manufacturers Association released a report with the following statistics: there are approximately 77.5 million dogs in the United States; 39 per cent of American households own at least one dog; and 24 per cent own two dogs. Nine per cent of households own three or more dogs, which is an average of 1.7 dogs for each household in America. According to the report, a staggering 75 per cent of all American dogs are neutered. The proportion of male to female dogs is the same, and 19 per cent of all dogs are adopted from an animal shelter. On average, American dog owners spend US$225 on vet bills annually.

DOG STORY

AUTHORS

BIOGRAPHIES
The Authors

Anna Pasternak
Jackie Stewart
Petronella Wyatt
Roy Hattersley
Annabel Goldsmith
Philip Treacy
Edward du Cann
Tom Rubython
Peter Egan

Anna Pasternak

Anna Pasternak is a leading British journalist and writes for *The Daily Mail*. She is the author of the popular 'Daisy Dooley Does Divorce' column which ran in *The Daily Mail* for four years and was turned into an international best seller. She currently writes the *On The Couch* column about her quest for love post the death of her beloved Dachshund Wilfred. She loved him more than any man, and, after losing him, wondered if she would ever feel a love like it again. She is the great niece of *Doctor Zhivago* author, Boris Pasternak.

Sir Jackie Stewart

Sir Jackie Stewart was Britain's most successful racing driver, winning three world championships between 1969 and 1973 before retiring at the age of 34 as world champion. In the past 37 years, he has forged an equally successful new career as a broadcaster, brand ambassador, race team owner and businessman. Now, aged 70, he still travels the world working for multinational corporations. His life has been beset with tragedy and death on the racing circuit; Jochen Rindt, François Cevert, Jim Clark and many others who died in horrible accidents were amongst his closest friends. But, in 2005, when his Norfolk Terrier 'Boss' died, he suffered a loss that caused him to re-assess grief altogether. He describes the experience as being equal to the mental anguish suffered in the loss of a close friend or family member.

Petronella Wyatt

Petronella Wyatt is the daughter of Lord and Lady Wyatt. Her late father was better known as Woodrow Wyatt, the prominent Labour politician and journalist. Latterly a conservative supporter, he became successful in business and was a leading socialite, counting the late Queen Mother and Baroness Thatcher amongst his very close friends. Petronella became a leading journalist and was formerly deputy editor of *The Spectator*. Now she writes for *The Daily Mail*. When her rare Papillon dog Mimi died, she hesitated to get a new dog, claiming Mimi was more like a human being and that no new puppy could compete with her memory.

Roy Hattersley

Roy Hattersley is a former Member of Parliament for Birmingham Sparkbrook. He was Deputy Leader of the Labour Party for nine years, from 1983 to 1992, and has been a Privy Councillor since 1975. Ennobled in 1993, he is still an active supporter and ambassador for the Labour Party. He is also an active journalist, most notably as a columnist for *The Daily Mail*, where he is now almost as well known for his love of dogs as he is for his politics. His relationship with his dog Buster, who went everywhere with him, transcended that with any previous canine companion. He openly admits that nothing has ever caused him as much pain as Buster's death.

Lady Annabel Goldsmith

Lady Annabel Goldsmith is the daughter of the eighth Marquess of Londonderry. She became more widely known by dint of her marriages to two of the most powerful and influential men in London during the seventies, eighties and nineties – Mark Birley and Sir James Goldsmith. Birley founded a nightclub in her name, called *Annabel's*, which was to become the forerunner of a nightclub empire. After divorcing Birley, she married Goldsmith, one of the great entrepreneurs of the 20th century. She watched him accumulate a two billion dollar fortune during their marriage. Her greatest success has been a wonderful family of five children (her eldest son Rupert died in 1986), all of whom are renowned personalities in their own right. But her most significant relationship was with a dog called Copper, whom she describes as the greatest character she ever knew.

Philip Treacy

Philip Treacy OBE is an Irish milliner and one of the 20th century's foremost hat designers. Since moving to England, he has designed hats for Alexander McQueen at Givenchy in Paris and for Karl Lagerfeld at Chanel. He runs his own eponymous label in London. He was born in 1967 in Ahascragh, a small village in County Galway in the west of Ireland and started sewing when he was five. He studied at the National College of Art and Design in Dublin and began designing and making hats as a hobby. His big break came in 1989, when his hats were shown to fashion director Michael Roberts and style director of *Tatler*, Isabella Blow. The following decade saw the business expand rapidly and he is now regarded as the world's top hat designer.

Sir Edward du Cann

The Rt Hon Sir Edward du Cann KBE was Conservative member of parliament for Taunton from 1956 to 1987 and served as chairman of the Conservative Party from 1965 to1967. He was best known as chairman of the powerful Conservative 1922 Committee for 12 years. He also served as Economic Secretary to the Treasury from 1962 to 1963 and as a Minister of State at the Board of Trade between 1963 and 1964. He was appointed as a Member of the Privy Council in 1964 and was chairman of the Public Accounts Committee from 1974 to 1979. Although a leading member of Edward Heath's shadow cabinet in the late sixties, he disagreed with Heath over Britain's prospective entry into the European Union and campaigned against it in the referendum to decide Britain's entry. As a result, he never served in Heath's cabinet, which played a leading role in overthrowing him as conservative party leader until the election of Margaret Thatcher in his place in 1975. But he rejected any further political office to continue chairing the 1922 Committee (the Conservative Parliamentary Party) for a record number of years, which he did until 1984 when he retired from front line political life and resigned his seat at the next election. He was also chairman of Lonrho, a very substantial company trading mainly in Africa, from 1987 to 1991, and is sometimes known as the founder of the modern Unit Trust movement. He was also MP for Taunton for a record number of years, and the founding Chairman of the Treasury Select Committee.

Tom Rubython

Tom Rubython is a well-known author specialising in the worlds of business and sport. Formerly editor of *BusinessAge*, *EuroBusiness* and *Formula 1 Magazine*, he also started the *Sunday Business* newspaper and, latterly, was editor of *BusinessF1* and *SportsPro* magazines. He is now a book publisher and author and was recently appointed editor of *Spectator Business* magazine. He lives with two blue roan Cocker Spaniels called Daisy and Doris. The oldest of the once-trio, Harriet (Harry), died earlier this year. Tom describes the finality he felt when Harry departed and attempts to find the reason why it hurt so much.

DOG STORY

AUTHORS

Peter Egan

Peter Egan is one of Britain's best known character actors, both on the stage and television. He came to fame in 1969 playing the gangster Hogarth in Granada's hit drama, *Big Breadwinner Hog*. He never looked back and, forty years later, he is still at the top of his profession. Peter Egan made his debut on British television in the BBC serialisation of *Cold Comfort Farm* in 1968. His other big successes include the role of Oscar Wilde in the series *Lillie* co-starring Francesca Annis. He also played the title role as future King George IV in the BBC series *Prince Regent* in 1979. Today, he regularly appears on the west end stage. He is married to actress Myra Frances and is a patron of the dog charity All Dogs Matter.

A

'A Dog's Prayer' 31
Ahascragh 161
Airedale Terrier 145
All Dogs Matter xiv, xxi,
    165
Akita 147
Alderney 105, 106, 107,
    110
Alfie 95
Alsatian 66
Althorp House 130
American Pet Products
        Mfers Assoc 153
Annabel's 160
Animal Welfare Bill 151
Aniston, Jennifer 152
Annis, Francesca 165
Antoinette, Marie 39
Ascot 24, 27
Autumn Cottage 122

B

Babylon 39
Ballet Rambert xvii
Balto 146
Banks, Trevor 16, 17
Barbour 135
Barnes, Arthur 66
Barney 66
Battersea 77, 95
BBC 86, 165
Bee 66
Begnins 20
Benjamin 70
Big Breadwinner Hog 165
Billing Wharf 116
Birley, Mark 160
Birley, Robin 66
Birmingham Sparkbrook
    159
Blair 145
Blossom 17, 20
Blow, Isabella xii, 91, 161
Blue Peter 149, 150
Bolsover 59

Border Collie xv, xiii, 66,
    103, 111, 128,
    145, 148, 150
Boss xi, xix, 15, 21, 22, 23,
    24, 25, 26, 27, 28,
    29, 32, 33, 157
Bourbon, Jacqueline 8, 10
Boxer 20, 101
Brighton 72
British Grand Prix 30
British Legion 106
Browning 37
Brown, Sheila 92
Bucharest 50
Buckingham Palace 105
Buckinghamshire 25
Budapest 50
Bugsy xi, 21, 22, 23, 24,
    25, 26, 27, 28, 30
Bull Mastiff 58
Bunhill Fields 127, 135
Burrell, Paul 67
BusinessAge 163
BusinessF1 163
Buster xi, xx, 52, 53, 54,
    55, 56, 57, 58, 60,
    62, 159
Buttons xii

C

Cagney, James, 38
Cann, Edward du xii, xx,
    100, 101, 162
Castle Ashby xii, 122, 135,
    136, 138
Cassie xiv
Cevert, François 157
Chance xii
Chanel 37, 50, 161
Chester Square 78
Chiswick High Road xvi
Cholmondeley, Marquis
    of 95
Christmas xii, 5, 59, 91,
    93, 135, 152
Christmas Eve 98
Clarence House 44

Clark, Jim 157
Clayton House 20, 22, 25,
    28, 29
Cleverly, Ann 134
Cocker Spaniel 129, 163
Cold Comfort Farm 165
Colefax and Fowler 74
Common Entrance Exam
    68
'Compassion' 145
Conservative 1922
        Committee 162
Conservative Party 162
Cooper 102
Cooper, Jilly 72
Copper xii, xx, 64, 68, 69,
    70, 71, 72, 73, 160
Corbett, David 149
Cornwall, Duchess of
    91, 98
County Galway 161
Couttisson, Amelia 92
Crackers xvi
Cruft's Dog Show 70
Custard xvii
Cyprus 107, 108, 110

D

Dachshund xx, 1, 6
Daisy 66, 113, 163
Dalmatian 84
Dee Dee xii
Delilah 134
Derbyshire 54, 59, 61
Design Museum 95
Dickens, Charles 143
Dinah 57
Dino xii
DJ xiii
Doctor Zhivago 156
Dolly 134
Doris xii, 113, 134
Dorsetshire 103
Duchess of Cornwall
    91, 98
Duke of Edinburgh 105,
    106

INDEX

Dushka 71
Dysart Arms 69

E

Earl Spencer 130
East Molesey 69
EasyJet 50
Eaton Square 85
Ebury Street 83, 84, 85
Edinburgh, Duke of 105
Edinburgh Book Festival
    xi
Edna xiii
Egan, Peter xi, xiii, xxi,
    165
Egham 24
Elizabeth Street 82, 86,
    89, 95, 97
Embankment 104
England 6, 23, 25, 39, 40
English Bull Terrier 58
EuroBusiness 163
Euston 125
Exmoor 150

F

Field Dog Trials 26
Fluffy 18
Fogle, Bruce 46
Formula One 23, 30
Formula 1 Magazine 163
Franca xiii
France 5, 6
Frances, Myra 165
Frankel, David 152
Frank, Scott 152
Ft Benton train station
    147
Fynn xiv

G

Gardner, Ava 44
Geldof 8
Givenchy 161
Goering 37
Goldsmith, Annabel xii,

xx, 65, 160
Goldsmith, James 160
Goldsmith, Zac 70
Grand Basset Griffon
    Vendeens 66
Grand Canal 39
Great Dane 102, 150
Great Serum Nome 146
Greco, Alessandra xx
Grogan, John 151
Guinness, Daphne xii

H

Hall, Jerry 94
Hambletonian 65
Ham Common 70
Hamilton, Lewis 50
Hampshire 37, 49
Hampstead Heath xvii
Hardy, Thomas 145
Harlech, Amanda xii
Harlestone xii, 124, 127
Harley 58
Harrison, Rex 51
Harrods 38
Harry the Hound xii, xxi,
    112, 113, 116,
    125, 129, 163
Harry's Bar 39
Hatfield 77
Hattersley, Roy xi, xx, 53,
    159
Heathrow Airport 23,
    96, 97
Helensburgh 19
Henley 2, 4
Hepburn, Audrey 51
Hepworth, Cecil 145
Herefordshire 72
Holland Park 92
Holly 135
Hollywood 7
Horserace Totalisator
    Board 44
Houghton 95
House of Commons 111
Houses of Parliament 104

I

Ireland 60, 161
Irish Club 85
Irish Terrier 145
Isabella 94, 95
Isle of Iona xi
Italian Grand Prix 124

J

Jack Russell xx, 83, 84
Jagger, Mick 94
Jaguar Racing 30
Jakie 58, 59, 62
Jane, India 71
Japan xx, 147
Jelfs, Yvette 83
Jessie 69
Jet Stream 31 22
John, Elton 87, 97
Jones, Grace xii, 82, 90, 97
Jules Rimet World Cup
    149
Juliet xii, xx, 100, 103,
    105, 106, 107,
    109, 111
Jumbone 132

K

Kennel Club 144
Khan, Jemima xii, 68, 70
Khan, Sulaiman 66
King Charles Spaniel 38,
    102
King Eystein of Norway
    141
King George IV 165
King Solomon 44
Kingston 69
Klemmie 39
Knightsbridge 67

L

Labrador Retriever xiv,
    xvi, xvii, 26, 151
La Bohème 40

Labrador 16, 20, 26, 28, 58, 128
Laika 148
Lagerfeld, Karl 161
Lassie Come Home 148
Liberal Party 150
Lily 66, 67
Lillie 165
Loch Lomond 15
London xvii, 4, 40, 54, 59, 61, 101, 125, 127, 128, 129, 131, 132, 133, 135, 142, 144, 160, 161
London Metropolitan Police 114
Londonderry, Marquess of 160
Lonrho 162
Lord Lundy 38
Lord Nelson 58
Lurcher Whippet xiv, 66, 145

M

Mail on Sunday XX
Mafia 72
Manolos 50
Marian 93
Marley & Me 37, 151, 152
Martin's Act 142
Megan xiv
McAlpine, Romilly 39
McMenamy, Kristin 93
McQueen, Alexander 161
Meldrum, Bill 21, 26, 31
Mimi xx, 40, 41, 42, 43, 44, 45, 47, 48, 49, 158
Minnie xi, xx, 38, 48, 49, 50, 51
Moët and Chandon 96
Moore, Henry 104
Moses 129
Moss, Karen xix
Moss, Kate 95

Mr Pig xii, xx, 76, 77, 83, 85, 86, 87, 88, 91, 92, 93, 94, 97, 98
My Fair Lady 51

N

Naish, Judith xx
Napoleon 38
National College of Art and Design 161
National Dog Club Committee 143
National Express coach 92
Newcastle-on-Tyne 143
Newton, Andrew 150
New York 151
New York Times 151
Nice 6
Ninochka xvii
Noakes, John 149, 150
Norfolk Terrier 15, 20, 21, 22, 33, 34, 66, 73, 157
Northamptonshire 125, 127

O

Old Shep 147
Old Street 125, 126
Old Street Police Station 114
Ormeley 65, 66, 73, 74
Oxfordshire 78

P

Papillon xx, 37, 39, 40, 49
Park House 122, 127, 128
Parliament 101
Passports for Pets 22
Pasternak, Anna xi, xx, 156
Pasternak, Daisy xi, xii, xx, 2, 4, 5, 6
Pasternak, Boris 156
Patch 149, 150

Peace Garden 65
Peak District 53, 61
Pearlstine, Maggie xx
Pease's Act 142
Pedigree Chum 45
Pekinese 20
Petra 149
Pickles 149
Pimms 33, 34
Platypus 71, 72
Plymouth 134
Pointer 105, 143
Pompadour, Madame de 39
Poodle 70, 71
Poppy 66
Portland Place 86
Premier League 61
Prince of Wales 19, 144
Prince Charles 91
Princess Diana 67, 130
Prince Regent 165
Promenade des Anglais 6
Pugh, Sylvia 82

Q

Queen Elizabeth 16, 17, 55, 60, 105, 106
Queen Mother 44, 45, 158
Queen Mother Hospital for Animals 77
Queen Victoria 143

R

Rainbow Bridge 11, 12
Rantzen, Esther 86, 87
Rawsthorne, Alice 95
Reservoir Dogs 91
Retriever 102
Rhodesian Ridgeback 23
Rhodes, Nick 94
Richmond 69
Richmond Park 69
Rimes, Tobey 146
Rigby & Peller 67
Rimet, Jules 149

INDEX

Rindt, Jochen 157
Rinker 150
Ripley's *Believe it or Not*
147
Ritchie, Guy 87
Riva 151
Roberts, Michael 161
Robot 148
Rocksavage, David 95
Romeo 103
Romilly 40
Roos, Don 152
Rottweilers 128
Royal British Legion 105
Royal Engineer Officer
105
Royal Parks 55
Royal Protection Group
17
RSPCA 142, 145, 146
Royal Veterinary College
77
Rubython, Tom xix, 113,
163
Rugby 21, 33
Rupert 65

S

Salty 151
Saluki 141
Sam xiv
Sandringham 20, 21
Sandringham Sydney 16
Saurr 141
Scott, Norman 150
Setter 143
Shaw, Andrew 67, 68
Sheepdog 145
Shibuya station 147
Siberian Huskies 146
Silverstone 30
Singleton, Valerie 149
Sky+ 4
Snare 16
SPCA 142
Somerset 101

Spaniels xiii, 16
Spectator Business 163
Spielberg, Steven 146
Spock, Mr 49
SportsPro 163
Springer Spaniel 19
Sputnik 2 148
Staffordshire XIV
Stewart Grand Prix 23
Stewart, Helen 16, 20, 22,
23, 24, 25, 26, 28,
29, 30, 32, 33
Stewart, Jackie xi, xix, 15,
157
St James's Park 55, 60
Stoppard, Tom 44
Stratford-on-Avon 124
Stubbs, John 21
Sunday Business 163
Sunday Times 83
Sunningdale 24, 25
Surbiton 69
Switzerland 20, 21, 22,
23, 26

T

Tatler 161
Taunton 162
Terriers 34, 73
Thatcher, Denis 43
Thatcher, Margaret 43,
158
The Belstone Fox 68
The Daily Mail 156, 159
The Falcon 136
The Times 66, 83
Thorpe, Jeremy 150
Tobey 146
Trace, Christopher 149
Treacy, Philip xii, xx, 77,
161
Treasury Select
Committee 162
Tucker xiv

U

UK 21, 23
USA 8, 20, 152
Ustinov, Peter 44

V

V&A 96
Venice 39
Victoria 93
Victoria coach station 92
Vienna 46, 47

W

Wallingford 92, 93
War Memorial 105
Warwickshire 143
Watson, Martin 24, 27
Wendel, Ella 146
Welsh Border Collie xiv
Westminster 104, 150
Westonbirt xi
Whisky 33, 34
Wilberforce, William 143
Wilde, Oscar xv, 165
Wilfred xi, xx, 1, 2, 4, 5, 6,
7, 10, 12, 156
Wilson, Owen 152
Wilson, Les xx
Wimbledon 59
Windsor 16, 21, 24
Winning Is Not Enough xix
Wodehouse, P.G. 37
Worcestershire 122
World Trade Centre 151
Wyatt, Petronella xi, 37,
158
Wyatt, Verushka xi
Wyatt, Woodrow 158
Wynyard 65

Y

Yag 94
Yorkshire 16

Photographic Credits

Cover	*Piotr Grzesik*
Foreword	*Sam Friedrich*
Chapter 1	*Anna Pasternak*
Chapter 2	*Mark Stewart Productions*
Chapter 3	*Les Wilson*
	Solo Syndication
Chapter 4	*Maggie Pearlstine Associates*
	Colin McPherson
	Corbis
Chapter 5	*Lady Annabel Goldsmith*
Chapter 6	*Philip Treacy/Alessandra Greco*
Chapter 7	*Sir Edward du Cann*
Chapter 8	*Tom Rubython/Graham Fudger*